KUHN'S INTELLECTUAL PATH

Thomas Kuhn's *The Structure of Scientific Revolutions* offers an insightful and engaging theory of science that speaks to scholars across many disciplines. Though initially widely misunderstood, it had a profound impact on the way intellectuals and educated laypeople thought about science. K. Brad Wray traces the influences on Kuhn as he wrote *Structure*, including his 'Aristotle epiphany', his interactions with James B. Conant, and his studies of the history of chemistry. Wray then considers the impact of *Structure* on the social sciences, on the history of science, and on the philosophy of science, where the problem of theory change has set the terms of contemporary realism/anti-realism debates. He examines Kuhn's frustrations with the Strong Programme sociologists' appropriations of his views, and debunks several popular claims about what influenced Kuhn as he wrote *Structure*. His book is a rich and comprehensive assessment of one of the most influential works in the modern sciences.

K. Brad Wray is Associate Professor at the Centre for Science Studies at Aarhus University, Denmark. He is the author of *Kuhn's Evolutionary Social Epistemology* (Cambridge University Press, 2011) and *Resisting Scientific Realism* (Cambridge University Press, 2018), and editor of *Interpreting Kuhn: Critical Essays* (Cambridge University Press, 2021).

KUHN'S INTELLECTUAL PATH

Charting *The Structure of Scientific Revolutions*

K. BRAD WRAY

Aarhus University

CAMBRIDGE
UNIVERSITY PRESS

CAMBRIDGE
UNIVERSITY PRESS

University Printing House, Cambridge CB2 8BS, United Kingdom

One Liberty Plaza, 20th Floor, New York, NY 10006, USA

477 Williamstown Road, Port Melbourne, VIC 3207, Australia

314–321, 3rd Floor, Plot 3, Splendor Forum, Jasola District Centre, New Delhi – 110025, India

103 Penang Road, #05–06/07, Visioncrest Commercial, Singapore 238467

Cambridge University Press is part of the University of Cambridge.

It furthers the University's mission by disseminating knowledge in the pursuit of education, learning, and research at the highest international levels of excellence.

www.cambridge.org
Information on this title: www.cambridge.org/9781316512173
DOI: 10.1017/9781009057882

First published 2021

A catalogue record for this publication is available from the British Library.

Library of Congress Cataloging-in-Publication Data
NAMES: Wray, K. Brad, 1963– author.
TITLE: Kuhn's intellectual path : charting The structure of scientific revolutions / K. Brad Wray.
DESCRIPTION: Cambridge ; New York, NY : Cambridge University Press, 2021. | Includes bibliographical references and index.
IDENTIFIERS: LCCN 2021016131 (print) | LCCN 2021016132 (ebook) | ISBN 9781316512173 (hardback) | ISBN 9781009057882 (ebook)
SUBJECTS: LCSH: Kuhn, Thomas S. | Kuhn, Thomas S. Structure of scientific revolutions. | Science – Philosophy. | Science – History. | Philosophers – United States – Biography. | Historians of science – United States – Biography. | BISAC: SCIENCE / Philosophy & Social Aspects
CLASSIFICATION: LCC Q143.K83 W73 2021 (print) | LCC Q143.K83 (ebook) | DDC 501/.092 [B]–dc23
LC record available at https://lccn.loc.gov/2021016131
LC ebook record available at https://lccn.loc.gov/2021016132

ISBN 978-1-316-51217-3 Hardback

For Lori

You know I would have been very glad to get a Nobel Prize – I certainly wanted fame in some sense.
—Thomas S. Kuhn, "A Discussion with Thomas S. Kuhn"

Contents

Acknowledgments

This book was written with the support of many people and institutions. The final draft of the manuscript was prepared while I was on sabbatical leave from Aarhus University in fall 2020. I had the pleasure of being a Visiting Fellow at Clare Hall, and a Visiting Scholar in the Department of History and Philosophy of Science, at the University of Cambridge. I thank Clare Hall, and the Department, especially Hasok Chang, my faculty sponsor, for this opportunity. While at Cambridge, I was also able to access Mary Hesse's papers at the Whipple Library. I am honored to have been recently elected a Life Member of Clare Hall. I also thank Paul Hoyningen-Huene, Stathis Psillos and Lutz Bornmann for supporting my applications to Clare Hall and the Department of History and Philosophy of Science.

I thank Aarhus University for the sabbatical leave, and in particular, the Dean's Office in the Faculty of Science for a sabbatical grant, which covered many of my costs related to relocating to Cambridge for the fall semester of 2020. I would also like to acknowledge the support of the Research Foundation at Aarhus University, Aarhus Universitets Forskningsfond (AUFF), for a starting grant I was awarded when I was first hired at AU – Grant: AUFF-E-2017-FLS-7–3. This grant has supported my research with Line Edslev Andersen on the epistemology of scientific publication, as well as other research-related costs for the past three years.

Parts of this book have been presented at a variety of conferences, where I received insightful feedback from the audiences. I thank my audiences from the following conferences: (i) the Estudios Kuhnianos: Pasado, Presente y Futuro conference at the Universidad de la República, in Montevideo, Uruguay, in November 2018; (ii) the Reviving Instrumentalism in Philosophy of Science conference at the Centre for Humanities Engaging Science and Society, Durham University, UK, in July 2019; (iii) the Methods and Problems in Philosophy of Science conference at the University of Castilla-La Mancha, Toledo, Spain, in September 2019; (iv) the Scientific Realism Workshop at the Centre for Science Studies, Aarhus University, Denmark, in October 2019; (v) the meeting of the Society for Realist/Antirealist Discussion at the Central Division Meeting of the American Philosophical Association, Chicago, in February 2020; and (vi) the Distinguished Lectures Series on Philosophy of Science, at Seoul National University, in November 2020. I presented papers based on drafts of two chapters of my book at the latter conference via Zoom. I also presented a part of the book at a colloquium at the Università degli Studi di Milano in January 2019, when I was visiting there as an examiner for a Ph.D. thesis defense.

The following individuals have given feedback in one way or another. I thank: Lori Nash, Nancy Cartwright, Darrell Rowbottom, Kyle Stanford, Julian Reiss, Omar El-Mawas, Robin Hendry, John Preston, Greg Frost-Arnold, Robert Northcott, Paul Hoyningen-Huene, Pablo Melogno, David Teira Serrano, Paula Atencia, Sam Schindler, Michael Devitt, Anjan Chakravartty, Otavio Bueno, Eric Oberheim, Howard Sankey, Leandro Giri, Hernán Miguel, Juan Vincente Mayoral, Lee McIntyre and Jamie Shaw. Paul Hoyningen-Huene and George Reisch have been especially helpful in guiding me to relevant documents in the Thomas S. Kuhn Archives at MIT. I also thank Jamie Shaw and Lori Nash for reading and commenting on the whole manuscript. Lori also prepared the index and corrected the proofs.

I acknowledge support from the following organizations: the British Society for the Philosophy of Science, the University of Castilla-La Mancha, the Faculty of Information and Communication at the University of the Republic in Montevideo, Uruguay, and the Argentinian Society of Philosophical Analysis (SADAF), and Aarhus Universitets Forskningsfond (Grant: AUFF-E-2017-FLS-7–3). I also thank the Centre for Science Studies at Aarhus University for their support.

I thank the referees for HOPOS for their constructive and thoughtful comments on earlier drafts of papers that I have either drawn on or integrated into the book.

Additionally, I am grateful to the following publishers for permission to reprint parts of the following articles that have been previously published in journals:

2013. "The Future of *The Structure of Scientific Revolutions,*" *Topoi: An International Review of Philosophy* 31:1, 75–79. Special issue on *Structure of Scientific Revolutions,* 50 Years On. DOI: 10.1007/s11245-012-9140-0

2015. "Kuhn's Social Epistemology and the Sociology of Science," in William J. Devlin and Alisa Bokulich (eds.), *Kuhn's* Structure of Scientific Revolutions – *50 Years On, Boston Studies in the Philosophy of Science*, Vol. 311, pp. 167–183. Dordrecht: Springer. DOI: 10.1007/978-3-319-13383-6_12

2016. "The Influence of James B. Conant on Kuhn's *Structure of Scientific Revolutions*," *HOPOS: The Journal of the International Society for the History of Philosophy of Science*, 6:1, 1–23. DOI: 10.1086/685542

2017. "Kuhn's Influence on the Social Sciences," in L. McIntyre and A. Rosenberg (eds.), *Routledge Companion to Philosophy of Social Science*, pp. 65–75. New York, NY, and London: Routledge.

2019. "Kuhn and the History of Science," in M. Fricker, P. J. Graham, D. Henderson, N. Pedersen and J. Wyatt (eds.), *Routledge Handbook of Social Epistemology*, pp. 40–48. New York, NY: Routledge.

2019. "Kuhn, the History of Chemistry, and the Philosophy of Science," *HOPOS: The Journal of the International Society for the History of Philosophy of Science* 9:1, 75–92.

2021. "Kuhn and the Contemporary Realism/Anti-Realism Debates," *HOPOS: The Journal of the International Society for the History of Philosophy of Science*, 11:1, 72–92.

I also thank the Department of Distinctive Collections at the MIT Libraries for their permission to cite material from their collections. I am especially delighted about the recent updating of their copyright permissions policy, which has streamlined the process significantly. I would also like to extend my thanks to Myles Crowley and Nora Murphy for providing a supportive and warm environment in which to conduct my research during my visits to the Department of Distinctive Collections. I am grateful to David Kaiser, of MIT, for providing me with some valuable information about the collection when I first visited it in 2013. David saved me a significant amount of time, enabling me to hit the ground running. I thank the State University of New York, Oswego, my former employer, for providing support for my research trips to Cambridge, Massachusetts. And I thank MIT and the Department of Linguistics and Philosophy for hosting me as a Visiting Scholar in fall 2015, when I was on a sabbatical leave in Cambridge, MA.

I thank Hilary Gaskin of Cambridge University Press for working with me on this project, and for supporting my work over the years. Thanks also go to Hal Churchman at Cambridge University Press, and I thank the referees for Cambridge University Press for their constructive feedback on and encouraging support of my project.

I dedicate this book to Lori Nash, my partner, who has both enjoyed and endured my career in philosophy, living in four different countries over the years. It is she who has had to reinvent herself each time we move, as well as learn the tax systems in each country.

Introduction

Thomas Kuhn's *The Structure of Scientific Revolutions* is a remarkable book. It has sold over 1 million copies, a startling number for an academic title.[1] It has been translated into numerous languages, ensuring that its reach is nearly endless. People have wondered how such a book could have come to be written.

Some have suggested that the success of *Structure* was quite accidental (see Fuller 2000), or that any number of other contemporary books could have achieved the success that *Structure* achieved. But I believe that there is another story to tell that explains why *Structure* is the book that it is and had the impact it had.[2]

Kuhn had quite atypical experiences, even before the publication of *Structure*. George Reisch has described Kuhn's unusual early education at a number of alternative progressive private schools in the United States (see Reisch 2019; see also Kuhn 1997/2000, 256–259). Kuhn suggests that these schools "made a major contribution to [his] independence of mind" (Kuhn 1997/2000, 257). Kuhn's experiences when he attended Harvard were also atypical. Though he was studying physics, he was able to foster his interests in both science and the humanities by writing for *The Harvard Crimson*, the student newspaper (Kuhn 1997/2000, 264; see also 268).[3] Kuhn completed his

---

[1] A letter from Penelope Kaiserlian, of the University of Chicago Press, from 1986, indicates that by 1986 the Press had sold 12,761 hard copies of the book and 633,924 paperback copies (see Kaiserlian 1986). This was ten years before the publication of the third edition.

[2] Steve Fuller (2000) compares Kuhn to Chance Gardiner, the protagonist in Jerzy Kosiński's *Being There* (see Kosiński 1970). Chance's unreflective remarks are misinterpreted by others as being profound. Thus, people are led to think of Chance as a very sophisticated and insightful thinker rather than the simple person that he is. Fuller believes that we have, similarly, inadvertently projected a profoundness on to Kuhn that is wholly undeserved. Further, Fuller suggests that our misunderstanding of what Kuhn wrote has shifted our thinking about science in quite radical ways, and not for the better.

[3] Further, as an undergraduate, Kuhn was also a member of the Signet Society, a very selective and prestigious "intellectual discussion society" at Harvard (see Kuhn 1997/2000, 268). In fact, he was the president of the society in his final year as an undergraduate. Incidentally, James B. Conant, who I will say more about shortly, had also been on the editorial board of *The Harvard Crimson* when he was a student (see Bartlett 1983, 92–93).

bachelor's degree in three years, and he immediately went to work at the Radio Research Laboratory, working for the American war effort (see Kuhn 1997/2000, 268–269). This is hardly the typical path even at Harvard.

When Kuhn returned from his service in the military, he began work on a dissertation in physics with John Van Vleck, who was then the Chair of the Physics Department at Harvard (see Bleaney 1982, 638). Kuhn was already familiar with Van Vleck, as he was the "head ... of the theory group" at the Radio Research Laboratory during the war (see Bleaney 1982, 628; 637; see also Kuhn 1997/2000, 268–269). Together with Van Vleck, Kuhn published "A Simplified Method of Computing the Cohesive Energies of Monovalent Metals," one of Kuhn's first publications. Van Vleck, in turn, was no ordinary scientist. He served as President of the American Physical Society in 1952–1953, and Vice President of the American Academy of Arts and Sciences in 1956–1957. And he would later go on to win the Nobel Prize for physics in 1977 (Bleaney 1982, 628).[4] Indeed, Van Vleck was not the only scientist of Nobel Prize caliber Kuhn encountered at Harvard. He also took courses in graduate school with Percy Bridgman and Julian Schwinger (see Kuhn 1997/2000, 267–268 and 274–275).[5]

In addition to working with a supervisor who was exceptional in his accomplishments, Kuhn also had peers who were exceptional. One of Kuhn's age peers was Philip W. Anderson. Anderson started his undergraduate education at Harvard in 1940, the same year Kuhn did, worked at the Radio Research Laboratory immediately after finishing his bachelor's degree, as Kuhn did, and returned to Harvard after the war to complete a Ph.D. under Van Vleck's direction, as Kuhn did (see Nobel 1977). Anderson won the Nobel Prize in Physics in 1977, along with Van Vleck

---

[4] It is interesting to note that Harriet Zuckerman discusses Van Vleck as an example of someone who occupies the forty-first chair (see Zuckerman 1977/1996, 158). The term "the 41st Chair" was coined by Robert K. Merton to describe the phenomenon that "a good number of scientists who have not received the [Nobel] prize and will not receive it have contributed as much to the advancement of science as some of the recipients, or more" (see Merton 1968/1973, 440). As Merton explains, "the phenomenon of the forty-first chair is an artefact of having a fixed number of places available at the summit of recognition" (Merton 1968/1973, 441). The term is derived from the situation created by the French Academy, where "only a cohort of forty could qualify as members" (see Merton 1968/1973, 441). Though Van Vleck won the Nobel Prize in Physics in 1977, Zuckerman's book was published that year, and it was obviously *written* without the knowledge that Van Vleck would win the prize in 1977. Zuckerman notes that Van Vleck, like other then-living occupants of the 41st chair, "[exhibits] the same pattern of early achievement, early recognition, and early institutional reward" (see Zuckerman 1977/1996, 158). For example, she notes that Van Vleck was awarded his doctorate when he was twenty-three years old and was "promoted to top academic rank while still in [his] twenties" (see Zuckerman 1977/1996, 158).
[5] Van Vleck had also taken courses with Bridgman (see Bleaney 1982, 632).

and Nevill Mott.[6] Kuhn was moving in elite circles, among people who were marked for success.

Kuhn worked with other exceptionally accomplished people during this time, most significantly, James B. Conant, the President of Harvard. The details of this relationship are outlined in Chapter 2, but I will highlight some of the key points here. When Kuhn was completing his Ph.D. in physics, Conant invited him to work as an assistant for him in the teaching of a General Education course in the history of science. This was part of an initiative of Conant's to ensure that the American elite were science-literate. It was a course designed for non-science majors. Conant's influence in America at this time was wide-ranging and profound. He had been an advisor on the Manhattan Project and played a significant role in the creation of the National Science Foundation. In fact, Conant was a widely known public figure, as he had appeared on the cover of *Time* magazine.[7] This involvement with Conant proved to be most significant, given the direction Kuhn's career subsequently took. In particular, Conant supported Kuhn's application to the Harvard Society of Fellows, which afforded Kuhn the opportunity to retrain as a historian of science upon completing his Ph.D. in physics (see Kuhn 1997/2000, 276; 278–279). Moreover, when Kuhn was writing *Structure*, he would draw extensively on material from the history of science course, specifically the cases from the history of chemistry that figured in the course. So Kuhn was, without a doubt, moving among some of the most powerful and accomplished researchers of his time.[8]

Although Kuhn was, like his peers, marked for success, his career was not without setbacks. He was denied tenure at Harvard in the mid-1950s, a significant psychological and professional blow. And he was denied a promotion to full professor in the Philosophy Department at UC Berkeley in the early 1960s, just before *Structure* was published.

Nonetheless, even through these challenging experiences, Kuhn's social capital as an academic continued to grow. The year *Structure* was published, he oversaw a project in Denmark sponsored by the American

---

[6] In a comprehensive study of American Nobel laureates, Harriet Zuckerman found that "more than half (forty-eight) of the ninety-two laureates who did their prize-winning research in the United States by 1972 had worked either as students, postdoctorates, or junior collaborators under older Nobel laureates" (see Zuckerman 1977/1996, 100).

[7] Conant would appear a total of four times on the cover of *Time* magazine, "'a rare record' for someone who had never held elected office" (Conant 2017, 478). The dates were: February 5, 1934, September 28, 1936, September 23, 1946 and finally September 14, 1959.

[8] This theme is stressed in Robert K. Merton's analysis of Kuhn's career in Merton's *The Sociology of Science: An Episodic Memoir* (see Merton 1977).

Physical Society, a project which involved constructing an archive of material related to the quantum revolution in physics in the early twentieth century. The goal of the project was to interview as many of the participants in the revolution as possible, before they died. This project put Kuhn in contact with many of the greatest physicists of the twentieth century. In fact, Kuhn interviewed Niels Bohr on a number of occasions, including just one week before his death.[9] Clearly, the fact that he was chosen to oversee such a project indicates that he was highly respected by physicists and historians of science, even before the publication of *Structure*.

As mentioned above, his interactions with Conant and involvement in the General Education science courses had a significant impact on Kuhn. The idea of writing a book about scientific revolutions first occurred to Kuhn in 1947, when he was preparing lectures for the course in the history of science. While trying to make sense of Aristotle's physics, he had a transformative experience. He came to realize that Aristotle was involved in a fundamentally different sort of enterprise than Galileo. And, rather than regarding Aristotle's worldview as mistaken, he saw that it provided a fundamentally different account of the world, that is, fundamentally different from either Galileo's or Newton's account. This experience also undermined his previous conviction that the growth of scientific knowledge is cumulative, with no significant setbacks.

But Kuhn's Aristotle epiphany was just the beginning of a very long process. As Kuhn notes, "it was fifteen years between the time these ideas *started* and the time [he] was finally able to write *Structure*" (see Kuhn 1997/2000, 292). Many other important insights that are central to *Structure* still

---

[9] Stanley Cavell provides another, more personal, sense of how remarkable Kuhn was:

> Kuhn was the product of two distinguished German Jewish families, accustomed to the best of everything in growing up and to being recognized for his intellectual accomplishments. When I told my uneducated father from Eastern Europe that not alone Kuhn but Kuhn's father had gone to Harvard, my father treated the news as something quite beyond comprehension. He repeated the words, as if searching for a history that could make them true. (Cavell 2010, 356)

Cavell and Kuhn were colleagues at UC Berkeley and continued to be friends long after both had moved back to the east coast, in the early 1960s. Kuhn had a great appreciation for Cavell. He notes that "the person [at Berkeley] who was *extraordinarily* important was Stanley Cavell. My interactions with him taught me a lot, encouraged me a lot, gave me certain ways of thinking about my problems, that were of a lot of importance" (see Kuhn 1997/2000, 297; emphasis in original). And in the Preface to *Structure*, in which Kuhn acknowledges his intellectual debts, he describes Cavell as "the only person with whom I have ever been able to explore my ideas in incomplete sentences. That mode of communication attests an understanding that has enabled him to point me the way through and around several major barriers encountered while preparing my first manuscript" (Kuhn 1962/2012, xlv–xlvi).

eluded Kuhn. Kuhn claims that his Lowell lectures, "The Quest for Physical Theory," given in 1951, were his first attempt to write *Structure* (see Kuhn 1997/2000, 289; Kuhn 1977, xvi). But, in the course of giving the lectures, he realized that he was not yet ready to write the book.[10] In fact, it was not until the mid-1950s that he began to really appreciate the role of normal science, the "periods governed by one or another traditional mode of practice ... [that] necessarily [intervene] between revolutions" (see Kuhn 1977, xvii). Without the concept of normal science, he was in no position to write the book.

Throughout his career, Kuhn would underestimate the amount of time and work it would take to complete projects. Robert K. Merton notes that

> by the age of thirty-two, when [Kuhn] made his application [for a Guggenheim Fellowship], he had published few articles: principally, one with Van Vleck in physics ... and the other, a historical piece on Boyle and structural chemistry in the seventeenth century which appeared in *Isis*. (Merton 1977, 91–92)

Kuhn's inability to estimate how long things would take, and the high standards that he held himself to, would repeatedly delay him in reaching his goals. As I have noted elsewhere, his lecture notes are marked up with critical remarks about how the lecture went, and what he would not do again next time (see Wray 2018b). In fact, when Kuhn died, he left an unfinished manuscript that he had been alluding to for decades.[11]

But the delays affecting the publication of *Structure* were not wholly detrimental. In fact, they allowed Kuhn to develop his ideas, and ensured that he did not publish the book prematurely. Some of the most influential concepts that figure in *Structure*, normal science and paradigm, for example, did not even enter Kuhn's mind until ten or so years after he first thought of writing a book about scientific revolutions. Kuhn had the good sense to wait until he had worked out his ideas. No doubt this is part of what explains the success of the book.

Structure was finally published in 1962. It was initially published as a volume in the *Encyclopedia of Unified Science*, a series that originated with the Vienna Circle positivists. As he was making the final revisions to

[10] In giving the lectures, Kuhn quickly recognized that he had not yet worked out his view adequately. As he explains, "the primary result of that venture was to convince me that I did not yet know either enough history or enough about my ideas to proceed toward publication" (Kuhn 1977a, xvi).

[11] Kuhn's final manuscript has a bit of a "pharaoh's curse" associated with it. Susan Abrams, the editor at University of Chicago Press with whom Kuhn was working before he died, died in 2003, and John Haugeland died in 2010. After Kuhn's own death, Haugeland was going to coedit the volume with James Conant, James B. Conant's grandson. It is still unpublished.

his manuscript, Kuhn expressed some concern to the publisher that his book would not get the attention it deserved, given the declining popularity of the *Encyclopedia* and its fading influence (see Kuhn 1997/2000, 300). In hindsight, we can say that the series was in its final days in the early 1960s. The particular volume that Kuhn was commissioned to write was intended to be devoted to the history of science. Its path to production was somewhat precarious. It seems that others had been invited to write the volume and had declined before Kuhn was invited to do so (see Kuhn 1997/ 2000, 291–292). Specifically, Kuhn notes that I. B. Cohen and Aldo Mieli had been asked to write the volume. When Cohen declined, he suggested Kuhn to the editors (see Kuhn 1997/2000, 292).

Kuhn's worries about publishing the book as a volume in the *Encyclopedia* were wholly unfounded. *Structure* was quickly regarded as a book well worth reading, not only by historians and philosophers of science, but also by many other academics and educated laypeople. Among Kuhn's collected papers, lectures and manuscripts in the archives at the Massachusetts Institute of Technology are countless letters. Many are from influential economists, psychologists and other academics. There is even a set of letters discussing a possible meeting between Kuhn and Newt Gingrich.[12] Many people were very excited by Kuhn's book.

In the 1970s, Merton noted that "in the first dozen years since its publication, [*Structure*] has given rise to *a library* of appreciative applications and diversely critical commentary" (see Merton 1977, 106; emphasis added). And since the 1970s this library has continued to expand. In fact, Kuhn was not only inundated with letters expressing positive responses to the book, but also manuscripts inspired by it. In 1973, in response to one fan who sent along a manuscript, Kuhn remarked that

> for better than five years I have been receiving two or three unsolicited manuscripts, sometimes book length, every week ... Though I very much hoped that my *Structure* would be widely read, I never dreamed of the nature or magnitude of the problems which its success would create for me. (Kuhn 1973)

Indeed, the many unsolicited manuscripts Kuhn received would not be the most significant problem that the book created for him. The most significant challenges he faced were the criticisms, many of them based on misunderstandings of *Structure*.

---

[12] See Thomas S. Kuhn Archives. MC240. Box 5: Folder 29, Congressional Clearinghouse.

My aim in this book is twofold. First, I aim to reconstruct the writing of *Structure*, clarifying the intellectual influences on Kuhn as he wrote the book. The existing studies of Kuhn have tended to focus on the influence of Kuhn's social milieu, understood in the broadest terms, with special attention given to the culture of Cold War America. Though these studies are often insightful, they fail to take adequate account of the intellectual influences on Kuhn as he wrote *Structure*. Second, I will trace the impact of *Structure*, with particular attention to its influence on the sociology of science, the history of science and the philosophy of science.

I will also discuss its broader influence, especially in the social sciences. In fact, Kuhn's influence in the social sciences is most interesting. Nowhere is the broad appeal of the concepts he developed more profound than in those fields. Social scientists found both the concepts and the general conception of science Kuhn developed highly fertile. The publication of *Structure* initiated a period of extensive reflection among social scientists on (i) the nature of the social sciences, (ii) their relationship to the natural sciences and (iii) the capacities of the social sciences to produce knowledge. This is quite ironic, as Kuhn claims to have discovered the paradigm concept while working among social scientists at the Center for Advanced Study in the Behavioral Sciences, at Stanford University. Kuhn describes how he realized that what the social sciences lacked, and what characterizes the natural sciences, are paradigms – fundamental research achievements that play an essential role in creating a consensus in a field and that make the sort of progress that we associate with the natural sciences possible. Kuhn also ignited a revolution of sorts in the sociology of science, one with which he was never fully comfortable.

I will also examine Kuhn's difficult relationship with the history of science. Though he spent many years working in history departments or involved in history of science programs, the impact he had on the history of science was comparatively insignificant: that is, when compared to the impact he had on the sociology of science and the philosophy of science. And recent assessments of his work by historians are not particularly flattering. Finally, I will also examine the impact Kuhn has had on the philosophy of science. On the one hand, Kuhn's legacy has been a set of problems that are a consequence of a particular reading of *Structure*, one that settled into place by around 1970. This has left us with a Kuhnian position of sorts that is widely deemed to be deeply problematic, as it threatens the integrity of science and scientific knowledge. On the other hand, his notion of revolutionary theory change left a lasting impact on debates and developments in contemporary philosophy of science, most notably in the realism/anti-realism debates. A central problem in these debates is what I call the problem of theory change, a problem that takes its form from Kuhn's *Structure*.

I will begin with a discussion of the Aristotle experience, the experience that set Kuhn on a new path away from a career in physics and toward a career in the history of science.

I have included information from Kuhn's curriculum vitae, prepared for an NSF application from the late 1980s, as a useful guide through the course of Kuhn's career.

Vita of Thomas S. Kuhn

(Source: TSK Archives, Box 20: Folder 12, NSF Research Reports)

Education

S.B. (summa cum laude), Physics, Harvard University, 1943
A.M., 1946
Ph.D., 1949

Positions Held

With radio research laboratory. Am-British Lab., OSRD, 1943–1945
Junior Fellow, Harvard Society of Fellows, 1948–51
Harvard University, 1951–56
   Assistant Professor, General Education and History of Science, 1952–56
University of California, Berkeley, 1956–1964
   Professor, History of Science, 1961–64
Princeton University, 1964–79
   M. Taylor Pine Professor of the History of Science, 1968–1979
Member, Institute for Advanced Studies, 1972–79
Fellow, New York Institute for the Humanities, 1978–79
Massachusetts Institute of Technology, 1979–
   Professor, Philosophy and History of Science, 1979–83
   Laurance S. Rockefeller Professor of Philosophy, 1983–

Honors and Fellowships

Guggenheim Fellow, 1954–55
Fellow, Center for Advanced Study in the Behavioral Sciences, 1958–59

PART I

The Groundwork for Structure*: Harvard*
1947 to 1955

I want to begin by exploring the origins of *Structure*, with special attention to the years Kuhn spent at Harvard after he returned from his time in the military. In Chapter 1, I examine Kuhn's Aristotle experience; that is, the experience he had when he read Aristotle's writings in physics, as he prepared some lectures as part of his contribution to the General Education science course he was working on with President James B. Conant. This is where it all began. That is, this is where Kuhn first decided that he would write a book on scientific revolutions. In Chapter 2, I provide a systematic analysis of James B. Conant's influence on Kuhn. It lays to rest a number of widely circulating claims about the inspiration for Kuhn's ideas. In Chapter 3, I provide an analysis of the influence of the history of chemistry on Kuhn's thinking. This chapter makes clear that Kuhn was concerned narrowly with the natural sciences, a point he insisted on repeatedly, despite the very broad appeal of the book. But it also draws attention to the fact that the histories of physics and astronomy were less significant in shaping Kuhn's views. Finally, in Chapter 4, I analyze the influence of the Logical Positivists on Kuhn's thinking during this period. Despite their significant influence in philosophy of science in America during the 1940s and 1950s, Kuhn does not engage directly with them in any significant way in *Structure*, and some have suggested that he had an outdated understanding of their views. I show that Kuhn engaged with the Logical Positivists more explicitly in the Lowell Lectures, his first attempt to write *Structure*. I also argue that insofar as he had a distorted picture of Logical Positivism, it was a consequence of W. V. Quine's influence.

CHAPTER I

What Did Aristotle Teach Kuhn?

Thomas Kuhn referred to his now-famous Aristotle experience on a number of occasions (see Kuhn 1977, xi–xii; 1987/2000, 15–20; Kuhn 1997/2000). And it is now commonplace for commentators of Kuhn's philosophy of science and history of science to discuss this incident, even if only in passing (see, for example, Bird 2000, 27; Fuller 2000, ch. 4, § 4; Andersen 2001, 2; Grandy 2003, 248; Nickles 2003, 144; Zammito 2004, 64; Hoyningen-Huene 2015, 194; Marcum 2015, 9–10; Kaiser 2016, 77; Reisch 2016, 13–17 and 24–26; Sankey 2018a, 82–83; Reisch 2019, 65–66 and 153–154; Burman 2020, 133–134, fn. 1). Indeed, so profound was the experience alleged to have been that it is not uncommon for it to be referred to as his Aristotle epiphany (see, for example, Reisch 2016, 16; and Heilbron 1998, 507).

My aim in this chapter is to examine the impact that this experience had on Kuhn's thinking, especially as he was writing *The Structure of Scientific Revolutions*. In many respects, this experience counts as one of the most profound influences on Kuhn as he wrote *Structure*. It rivals both (i) his experience working with James B. Conant on the General Education science courses at Harvard, and (ii) the year he spent at the Center for Advanced Study in the Behavioral Sciences, where he discovered the importance of paradigms for *natural* scientists, and their absence in the social sciences (see Kuhn 1962/2012, xlii). As we will see, the Aristotle experience was the source of Kuhn's initial discovery of scientific revolutions, that is, those disruptive changes in science that undermine the strictly cumulative account of scientific progress that he reacted against in *Structure* (see Kuhn 1977, xiii). That experience thus marks the beginning of his long journey toward writing *The Structure of Scientific Revolutions*.

I will also identify key parts of Kuhn's project that were not yet within his grasp in 1947, when he had the Aristotle experience. I thus explain why Kuhn was in no position to complete a book like *Structure* then. Indeed, key ingredients would elude his grasp for years. So, my aim here is to understand how much Aristotle taught Kuhn, and how much Kuhn would need to learn before he could write *Structure*.

What Happened in the Summer of 1947?

Interestingly, Kuhn does not discuss the Aristotle experience in either *The Copernican Revolution* or *The Structure of Scientific Revolutions*, his first two books. Rather, the first sustained published discussion of the experience did not take place until 1977, in the Preface to his collection of papers, *The Essential Tension*. The discussion of the experience there runs for three pages. It is worth examining this account in detail.

We can begin with Kuhn's account of the facts. The experience happened in the summer of 1947 (see Kuhn 1977, xi–xii). He was preparing "a set of lectures on the origins of seventeenth-century mechanics" (Kuhn 1977, xi). These lectures were to be part of his contribution to the General Education Natural Science course he had been invited to work on with Conant. As Kuhn explains, in order to prepare his lectures on the origins of seventeenth-century mechanics, he felt he "needed first to discover what the predecessors of Galileo and Newton had known about the subject, and preliminary inquiries soon led [him] to the discussion of motion in Aristotle's *Physica* and to some later works descending from it" (Kuhn 1977, xi). Kuhn had initially thought that Galileo would have built on the work of Aristotle and contemporary Aristotelians. Scientific knowledge, he had assumed, was more or less cumulative in its growth.

But this is not what he found. Rather he was startled, largely as a consequence of what he unreflectively brought to his reading of the texts. As Kuhn explains, he "approached these texts knowing what Newtonian physics and mechanics were" (Kuhn 1977, xi). As a consequence of his own immersion in Newtonian physics, he reports that he approached the texts with the following two questions in mind: "(1) How much mechanics was known within the Aristotelian tradition, and (2) how much was left for seventeenth-century scientists to discover?" (Kuhn 1977, xi; numerals added). So he had approached the texts assuming that Aristotle's project was more or less the same as Galileo's and Newton's. Aristotle, Galileo and Newton were all physicists, after all, or so Kuhn had thought.

Initially, Kuhn was quite perplexed by the extent of Aristotle's ignorance of *mechanics*. As he explains, "the more I read, the more puzzled I became. Aristotle could, of course, have been wrong – I had no doubt that he was – but was it conceivable that his errors had been so blatant?" (Kuhn 1977, xi). Kuhn found it almost impossible to believe that Aristotle had been so misguided in his understanding of mechanics. The persistence of Aristotle's influence through the ages seemed irreconcilable with such a view.

Then the epiphany happened. Kuhn reports that "those perplexities *suddenly* vanished" (xi; emphasis added). Elaborating, Kuhn goes on to say that "I *all at once* perceived the connected rudiments of an alternative way of reading the texts with which I had been struggling" (Kuhn 1977, xi; emphasis added). His understanding of Aristotle and of Aristotle's writings was transformed. He felt that he could *now* understand the Aristotelian worldview. Kuhn's description of his transformative experience sounds much like the sort of shift one experiences when one sees the second image in a Gestalt figure, that is, the old lady as well as the young lady, or the duck as well as the rabbit, after a period of initially not seeing it.

What had changed? According to Kuhn, "for the first time [he] gave due weight to the fact that Aristotle's subject was *change-of-quality in general*, including both the fall of a stone and the growth of a child to adulthood" (Kuhn 1977, xi; emphasis added). He now recognized that Aristotle's *physics* was not principally a *science of mechanics*. Thus, he realized that Aristotle was not even engaged in the same enterprise as Galileo and Newton, even though we are accustomed to tracing a lineage from Aristotle to Galileo, and then to Newton.

As a consequence of Kuhn's profound change in understanding, a number of other aspects of Aristotle's physics fell into place for him. First, Kuhn realized that mechanics was not a central part of Aristotle's concerns. For Aristotle "*the subject* that was to become mechanics was at best a still-not-quite-isolable special case" (Kuhn 1977, xi; emphasis mine). Generalizing from this case, Kuhn learned that scientific fields are not fixed by subject matter once and for all. In time, Kuhn would emphasize the importance of not using contemporary terms, like physics, for example, to refer to earlier scientific fields and practices from which these contemporary practices evolved (see Kuhn 1977, xv–xvi; see also Kuhn 1997/2000, 290 and 295). This, he thought, might aid historians in not projecting back on to earlier scientific practitioners interests and concerns that had not been part of theirs.[1] This underscores his conviction that theories in a field

---

[1] Indeed, Kuhn would later express regret that he had failed to take this precaution in *Structure* (see Kuhn 1997/2000, 290 and 295). Kuhn claims that this is a common defect of histories of science written by scientist-historians (see Kuhn 1971/1977, 149). In fact, he notes that "sometimes a specialty

are not profitably conceived as successive attempts to get at the same underlying reality. That is, the history of a scientific field is not fruitfully told as a history of the convergence on a fixed reality.[2]

Second, Kuhn also realized that the Aristotelian ontology was fundamentally different from the Newtonian ontology.

> The *permanent ingredients* of Aristotle's universe, its ontologically primary and indestructible elements, were not material bodies but rather the qualities which, when imposed on some portion of omnipresent neutral matter, constituted an individual material body or substance. (Kuhn 1977, xii; emphasis mine)

Further, Kuhn claims that he realized that "position itself was . . . a quality in Aristotle's physics" (xiii). And he also realized that "in a universe where qualities were primary, motion was necessarily a change-of-state rather than a state" (xii). By virtue of these ontological differences between Aristotle's theory and Newton's theory, he came to believe that Aristotle and Newton had lived and worked in different worlds. In John Heilbron's apt phrasing, Kuhn realized that "Aristotle had not been writing bad Newtonian physics but good Greek philosophy" (Heilbron 1998, 507). Further, as Paul Hoyningen-Huene notes, with this experience Kuhn "caught a glimpse of incommensurability," a notion that would come to play a significant role in *Structure* (see Hoyningen-Huene 2015, 194).

Third, Kuhn was now able to appreciate *the integrity of the Aristotelian worldview*. For example, Kuhn explains that "the exposure to Aristotle . . . taught [him] the integrity [of Aristotle's] quadripartite analysis of causes" (Kuhn 1977, xiv). More generally, he learned that earlier theories had their own integrity, an integrity that can often only be appreciated if one recognizes that earlier scientists were not aiming to do what those that followed aimed to do. The differences between them and their successors are not signs of failure, but rather indicate different concerns. Indeed, here we have the basis for Kuhn's view that successive theories in a field are

---

which they traced from antiquity had not existed as a recognized subject of study until a generation before they wrote" (Kuhn 1971/1977, 149).

[2] According to Kuhn, the history of science presented in science textbooks is built on such an assumption. As Kuhn explains, "partly by selection and partly by distortion, the scientists of earlier ages are implicitly represented as having worked upon the same set of fixed problems and in accordance with the same set of fixed canons that the most recent revolution in scientific theory and method has made seem scientific" (Kuhn 1962/2012, 137). In the context of science textbooks, this distortion, Kuhn suggests, may serve an important function, perhaps motivating students (see Kuhn 1962/2012, 136–137 and 164–165).

fundamentally different and cannot be aptly described as a series of ever more accurate approximations converging on the truth.

With the perspective gained from this transformative experience, Kuhn realized that he had inadvertently approached Aristotle's texts with pre-suppositions that were an impediment to his understanding them and Aristotle's project in general. As Kuhn explains, "being posed in a Newtonian vocabulary, [the] questions [he asked] demanded answers in the same terms, and the answers then were very clear" (Kuhn 1977, xi). As Kuhn reports, "even at the apparently descriptive level, the Aristotelians had known little of mechanics; much of what they had had to say about it was simply wrong. No such tradition could have provided a foundation for the work of Galileo and his contemporaries" (Kuhn 1977, xi).

But after he discovered "a new way to read a set of texts," Kuhn reports that he "had few problems understanding why Aristotle had said what he did about motion or why his statements had been taken so seriously" (Kuhn 1977, xii). As Kuhn explains, though he "did not become an Aristotelian physicist as a result ... [he] had to some extent learned to think like one" (xii). That is, Kuhn could now see the world as Aristotle had, and understand why the sorts of research problems that had engaged him and his followers seemed important. Indeed, this aspect of historical scholarship would continue to intrigue Kuhn (see Kuhn 1997/2000, 280).

Kuhn describes the experience as his "own enlightenment." And most significantly, it led to his discovery of *the nature of scientific revolutions* (Kuhn 1977, xi). As Kuhn explains,

> what [his] reading of Aristotle seemed ... to disclose was a global sort of change in a way men viewed nature and applied language to it, one that could not properly be described as constituted by additions to knowledge or by the mere piecemeal correction of mistakes. (Kuhn 1977, xiii)

This is the key insight Kuhn gained from the Aristotle experience, and this is why he claims the project that culminated in the publication of *Structure* began in 1947 (see Kuhn 1962/2012, xxxix). Scientific revolutions were central to Kuhn's new understanding of the growth of scientific knowledge.

Kuhn does draw some additional lessons from the experience, lessons concerning historiography and pedagogy. For example, Kuhn claims that, from this experience and similar ones with other classic texts in the history of science, he learned that "there are many ways to read a text, and the ones most accessible to a modern are often inappropriate when applied to the past" (Kuhn 1977, xii). Thus, he learned that the past is, at least initially, in some sense opaque to us, especially in those cases where we must, as historians of science, enter into a radically different theoretical framework and worldview.

But he also claims that the opacity of radical theoretical frameworks is not an insurmountable problem. He explains that "the plasticity of texts does not place all ways of reading on a par, for some of them (ultimately, one hopes, only one) possess a plausibility and coherence absent from others" (Kuhn 1977, xii). Thus, Kuhn did think that there were historical facts that we could in principle get at, or at least aim to get at. In no way was Kuhn led to some sort of unconstrained relativism about the history of science.

Finally, Kuhn gives students a heuristic to guide them in their endeavors to understand the scientific past, one that he learned from the Aristotle experience: "When reading the works of an important thinker, look first for the apparent absurdities in the text and then ask yourself how a sensible person could have written them" (Kuhn 1977, xii). These passages, he suggests, are the key to a more authentic reading of the texts. But Kuhn claims that "when those passages make sense, then you will find that more central passages, ones you previously thought you understood, have changed their meaning" (Kuhn 1977, xii). Thus, the process of making sense of theoretical frameworks and scientific practices from the past will change what we thought we understood. It is a hermeneutical process (see Kuhn 1977, xiii).

Kuhn discussed the Aristotle experience again in his 1987 paper "What Are Scientific Revolutions?" By this time, forty years had passed since it had taken place. The recounting of the story is quite similar to his account from 1977; not surprisingly, given that he could have reread his own earlier account as he set out to retell the story. But he does add some details. For example, see Kuhn's description of the circumstances of the experience: "I was sitting at my desk with the text of Aristotle's *Physics* open in front of me and with a four-colored pencil in my hand. Looking up, I gazed abstract-edly out the window of my room – the visual image is one I still retain" (Kuhn 1987/2000, 16).[3] Next, he explains that "suddenly the fragments in my head sorted themselves out in a new way, and fell into place together. My jaw dropped, for all at once Aristotle seemed a very good physicist

---

[3] Here Kuhn seems to employ a trope common in ethnography, recounting his experience in detail in order to give credibility to what he is about to tell us, just as the ethnographer gives a detailed description of what they see and experience as they arrive at the location where they will do their fieldwork.

indeed, but of a sort I'd never dreamed possible" (1987/2000, 16). The key here is his achieving a hitherto unanticipated new understanding.

In 1987, Kuhn makes clear some of the general lessons about scientific revolutions he drew from the experience, lessons that he had only alluded to in the earlier telling. Kuhn explains that "that sort of experience – the pieces suddenly sorting themselves out and coming together in a new way – is . . . [a] general characteristic of revolutionary change" (Kuhn 1987/2000, 17). With scientific revolutions, Kuhn notes, "the central change cannot be experienced piecemeal, one step at a time. Instead it involves some relatively sudden and unstructured transformation in which some part of the flux of experience sorts itself out differently and displays patterns that were not visible before" (Kuhn 1987/2000, 17). Kuhn thus draws attention to the *holistic* nature of the change from one theory to another. In this respect, his appeal in *Structure* to the duck/rabbit image is apt. One sees either a duck or a rabbit. There is no transitional figure open to our perception, something that is part duck and yet also part rabbit.

After describing some of the details of Aristotle's physics, Kuhn notes that

> those remarks . . . should sufficiently illustrate the way in which Aristotelian physics cuts up and describes the phenomenal world. Also, and more important, they should indicate how the pieces of that description lock together to form an integral whole, one that had to be broken and reformed on the road to Newtonian mechanics. (Kuhn 1987/2000, 20)

Kuhn makes explicit the holistic nature of theories, and the implications this has for both (i) the way they are understood, and, ultimately, (ii) the way they are overthrown. To learn a theory one must learn a cluster of concepts together. And a scientific revolution, unlike normal scientific research, involves a radical reworking of the scientists' worldview. This is why scientific revolutions are disruptive and are experienced as such by the scientists involved.

Kuhn also makes some passing remarks about the Aristotle experience in the interview published in *The Road since Structure*; this interview was conducted in 1995, near the end of his life (see Kuhn 1997/2000, 275, 276, 278, 285 and 292–293). Some of these remarks add no details beyond what he had already reported earlier. He does, though, note how excited he was to be given the opportunity to work with Conant on the course, and his surprise at being asked to "go out and do a case study on history of mechanics for this course" (Kuhn 1997/2000, 275). Kuhn also suggests that it was after the first semester of teaching the course with Conant that he realized he wanted to change his career path. As he explains, "I wanted to teach myself enough history of science to establish myself there in order to do the philosophy," for ultimately, Kuhn's ambitions were philosophical (see Kuhn 1997/2000, 276). The intended audience for the book he wanted to write about scientific revolutions was philosophers of science (see Kuhn 1962/2012, xxxix–xl).

Kuhn wanted to write a book about scientific revolutions in order to show how key discoveries in science can only be made by working with a radically new theory, one that makes assumptions about the world that are fundamentally different from the assumptions made by the replaced theory. This was the key insight Kuhn had gained that summer. And this was an important catalyst for his idea for a book on science, the book that would ultimately become *The Structure of Scientific Revolutions*. In fact, Kuhn repeatedly remarked that he "had wanted to write *The Structure of Scientific Revolutions* ever since the Aristotle experience" (Kuhn 1997/2000, 292; see also Kuhn 1977, x; but also Kuhn 1962/2012, xxxix).

What Kuhn Still Had to Learn

So far, I have argued that the Aristotle experience had taught Kuhn about the *nature* of scientific revolutions. Importantly, Kuhn came to understand that different theories in a scientific field have their own integrity and that the succession of theories in a field cannot be appropriately described as steps that bring us ever closer toward a final true theory. Rather, earlier theories, when understood correctly, in *their* cultural context, provided the means for progressive scientific practices and traditions. Aristotle's theory served the interests of his contemporaries in ways similar to those in which Newton's theory served the research interests of his contemporaries. And each new theory in a scientific field is not aptly described as aiming at the same things as the theory it replaced. Rather, the *field* itself shifts and changes with each change of theory, making claims of progress *through* revolutionary changes of theory somewhat problematic (see Kuhn 1997/2000, 292). These were important insights for Kuhn, and they would play a central role in his theory of science, and thus figure importantly in *Structure*, when he was ready to write it. But Kuhn was still a long way off from being able to write *Structure* after the summer of 1947.[4] There are

---

[4] Galison notes that Kuhn had sketched the outline of a book in a notebook from 1949. Here the proposed title was "The Process of Physical Science" (see Galison 2016, 55–56). This was even before the Lowell lectures, which Kuhn described as his first attempt to write *Structure*.

three important things about scientific revolutions that he had not yet grasped, and that would delay the writing of *Structure*. In fact, fifteen years would elapse between the Aristotle experience and the publication of *Structure*.

First, in 1947 Kuhn had not yet grasped the role or function of scientific revolutions in the growth of scientific knowledge. He knew that scientific revolutions threatened the cumulative account of the growth of scientific knowledge. But, in *Structure* Kuhn is quite clear that revolutions play a significant and definite function in the growth of scientific knowledge. Because every scientific theory ultimately encounters anomalies that expose its limitations, revolutionary changes of theory are an integral part of the growth of scientific knowledge. As Lydia Patton astutely notes, "it is an axiom of Kuhn's account that no paradigm can deal with all the phenomena" (see Patton 2018, 116). Revolutionary changes of theory are scientists' means of normalizing the anomalies that a theory brings to light, but cannot adequately resolve or normalize. That is, revolutionary changes of theory are scientists' way of adapting once they have run up against the limits of the theory they have been working with. Melogno and Courtoisie express the point as follows: "scientific revolutions constitute a key mechanism for scientific progress" (Melogno and Courtoisie 2019, 27). In 1947, Kuhn had not yet realized this.

Second, Kuhn had not yet put scientific revolutions into the general schema that emerges in *Structure*, the cyclical pattern of change that he claims characterizes the growth of knowledge in the natural sciences. This is not surprising as revolutions are only part of the cyclical pattern. He still lacked the necessary ingredients to articulate what role revolutions play in science, even if he did recognize that such revolutions exist.

Third, by 1947, Kuhn was in no position to describe or account for the periods of science between revolutionary changes of theory. That is, he had not yet conceived of normal science, the tradition-bound research activities that occur between revolutions. Kuhn suggests that this piece of the puzzle did not fall into place until the late 1950s. In fact, he suggests that it was while he was working on a late draft of his paper "The Function of Measurement in Modern Physical Science" that he began to work out the nature of normal science (see Kuhn 1977, xvii). He explains that in a late revision of that paper he introduced a section titled "Motives for Normal Measurement," where he claims "the bulk of scientific practice is . . . a complex and consuming *mopping-up operation* that consolidates the ground made available by the most recent theoretical breakthroughs" (see Kuhn 1977, xvii; see also Kuhn 1997/ 2000, 295; emphasis added).[5]

Kuhn's paper on measurement is devoted to debunking the idea that measurements in science are driven by the desire to either test or confirm theories, a view that Kuhn felt was implied by scientific textbooks. Indeed, measurement often involves the extension of theory, a mopping up of sorts. This notion of mopping up would make its way into *Structure* (see Kuhn 1962/2012, 24); it was crucial, as it made sense of what happens between revolutions, when a theory is taken for granted and assumed to provide an accurate description of the world.

In fact, the elusiveness of normal science would prove to be a real barrier for Kuhn in his efforts to write *Structure*. Kuhn notes how challenging it was initially for him to write on normal science to work out his ideas, even as late as 1958 when he was at the Center for Advanced Study in the Behavioral Sciences (see Kuhn 1997/2000, 296). He was still stuck in a particular framework that was proving rather unfruitful. As he explains, he "was taking a relatively classical, received view approach to what a scientific theory was – . . . [attributing] all sorts of agreement about this and that, and the other thing, which would have appeared in the axioma- tization either as axioms or as definitions" (Kuhn 1997/2000, 296). Thus, Kuhn was still under the spell of the Logical Positivists, to some extent. With the aid of the "paradigm concept," specifically, the notion that much of the consensus in scientific research communities is on exemplars or models, rather than theories, which are expressed explicitly in propositions, things fell into place (see Kuhn 1997/2000, 296). The successful conduct of normal science requires the aid of exemplars, as well as a theoretical framework. And this insight enabled Kuhn to abandon the conception of scientific theories with which he had been working, a conception that was strongly influenced by Logical Positivism.

The potent combination of (i) the paradigm concept and (ii) the notion of mopping up enabled Kuhn to resolve the issue to his satisfaction. He now understood what scientists were doing between scientific revolutions.[6] And this enabled him to clarify the function of scientific revolutions, setting them in the context of the cycle of change. None of this, though, emerged from the Aristotle experience in the summer of 1947. Indeed, as

---

[5] We can be even more precise here, if Kuhn's memory is to be trusted. He claims to have completed the revised draft that integrated this section in the spring of 1958 (see Kuhn 1977, xvii).

[6] The elusiveness of normal science, that is, the practice of science between revolutionary changes of theory, caught Karl Popper off guard as well. Indeed, though critical of the practice of normal science, Popper acknowledges that Kuhn drew his attention to the notion (see Popper 1970/1972).

late as 1958, Kuhn was still working this out. Thus, George Reisch is mistaken in claiming that "by 1951 ... *Structure*'s philosophy of science was largely in place" (Reisch 2016, 18).[7] There was still much to work out even in 1951.

What About the Historians Kuhn Was Reading?

It is worth briefly examining the influence that the various historians whom Kuhn was reading had on his thinking in this early period. In particular, one might be led to hypothesize that it was their influences that led him to the correct reading of Aristotle, the one that led Kuhn to recognize that science progresses through revolutionary changes of theory.

In the Preface to *Structure*, for example, Kuhn identifies a group of historians who were especially influential on his thinking. Specifically, he identifies a group of French historians: Alexandre Koyré, Emile Meyerson, Hélène Metzger and Anneliese Maier (see Kuhn 1962/2012, xl). He also mentions Arthur Lovejoy: specifically, his *Great Chain of Being* (see Kuhn 1962/2012, xl). In the Preface, Kuhn claims that "their works ... have been second only to primary source materials in shaping my conception of what the history of scientific ideas can be" (Kuhn 1962/2012, xl). He also mentions Ludwik Fleck in the Preface, specifically Fleck's "almost unknown monograph, *Entstehung und Entwicklung einer wissenschaftlichen Tatsache*" (see Kuhn 1962/2012, xli). Is it possible that these historians of science drew Kuhn's attention to the significance of scientific revolutions?

I do not think so. In fact, there is good reason to believe that these historians were not the source of Kuhn's ideas about scientific revolutions, that is, the ideas that set him on course to writing *Structure*. After all, Kuhn notes that he only encountered the work of Meyerson, Metzger, Maier and Fleck when he was a junior fellow at the Society of Fellows at Harvard. That would mean that he only encountered them in or after 1949. That is two years after the Aristotle experience and Kuhn's experience working with Conant on the General Education Natural Science course. Indeed,

---

[7] Reisch proceeds to list the various things that were "in place" for Kuhn by 1951. They include the following: "[1] experience underdetermines theory, [2] theory and observation were dependent and 'intermingled,' [3] theories were understood as holistic sets of ideas or conceptual schemes, and [4] the scientific mind was unaware that it operates within only one possible system of ideas ... that in *Structure* would lead Kuhn to characterize scientific revolutions as 'invisible' to most scientists" (Reisch 2016, 18; numerals added). Oddly, shortly afterwards, Reisch claims that "what was missing at this early stage ... was *Kuhn's theory of paradigms*" (see Reisch 2016, 18; emphasis added). This seems irreconcilable with Reisch's claims that "by 1951 ... *Structure*'s philosophy of science was largely in place" (Reisch 2016, 18).

Kuhn states explicitly that it was only in 1950 that he was made aware of Meyerson's work. In a discussion with Karl Popper, when Popper was giving the William James Lectures at Harvard, Popper drew Meyerson's work to Kuhn's attention (see Kuhn 1997/2000, 286–287).[8]

It is interesting to see how Kuhn describes the influence of Fleck's book. Kuhn claims that Fleck's "essay ... anticipates many of my own ideas" (Kuhn 1962/2012, xli). But, elaborating, Kuhn notes that "Fleck's work made me realize that those ideas might require to be set in *the sociology of the scientific community*" (Kuhn 1962/2012, xli; emphasis added). So, if we are to take Kuhn's account as accurate, it was the sociological dimensions in Fleck's work, not the historical dimensions, that left their mark on Kuhn. Kuhn, though, was uncomfortable with the specific details of Fleck's "sociology." In particular, Kuhn claims that he "never felt at all comfortable ... with [Fleck's] 'thought collective.' It was clear it was a group, since it was a collective, but [Fleck's] model ... was the mind and the individual" (Kuhn 1997/2000, 283). That is, Fleck was ascribing properties that we associate with individuals to scientific collectives in a manner that struck Kuhn as implausible.

In his Foreword to the English translation of Fleck's *Genesis and Development of a Scientific Fact*, Kuhn explains his chief problem with Fleck's analysis in more detail. In Kuhn's words:

> what troubles me is [that] ... the notion [of a thought collective is] intrinsically misleading and a source of recurrent tensions in Fleck's text. Put briefly, a thought collective seems to function as an individual mind writ large because many people possess it (or are possessed by it). To explain its apparent legislative authority, Fleck ... repeatedly resorts to terms borrowed from discourse about individuals. (Kuhn 1979, x)

When Kuhn developed his own "sociology of science" he assumed that scientific research communities, the sorts of groups that work with a theory and undergo revolutionary changes of theory, were composed of many individuals, each different from the others in subtle but important ways. These individual differences would play a crucial role in Kuhn's account of scientific change. The groups, as far as Kuhn was concerned, were like biological species, where individual differences really matter. They play a crucial role in understanding how the community as a whole responds to

---

[8] Kuhn claims that he "didn't like the philosophy at all" in Meyerson's *Identity and Reality*, "but, boy, did [he] like the sorts of things [Meyerson] saw in historical material ... [Meyerson] was getting it right in ways that were different from the ways that history of science was being written" (Kuhn 1997/2000, 287).

a new challenge, be it a change in the environment, in the biological case, or an anomaly, in the scientific case.

Koyré is a more complicated case. Kuhn did read Koyré's work as he prepared the material on the history of mechanics for the course with Conant. In particular, Kuhn read *Études Galiléennes* (see Kuhn 1990/2016, 21). And he often encouraged students to read Koyré's work. Indeed, John Schuster, who was a Ph.D. student in history at Princeton when Kuhn taught there, recalls that "when Kuhn addressed each year's crop of new history of science graduate students, he would make a point of bringing in his well-worn, pre–World War II copy of Koyré's *Études galiléennes* . . . He would intone, 'Nobody is leaving here until they have read all of this'" (Schuster 2018, 395, fn. 7). Koyré's work was thus really important for Kuhn. But nothing like the picture of science that we find in *Structure* can be found in Koyré's work. The type of influence Koyré could have had on Kuhn would have been methodological, specifically an internalist approach to the history of science. In fact, in a paper on the relationship between the history of science and the philosophy of science, Kuhn reports that "from [Lovejoy and Koyré] my colleagues and I learned to recognize the structure and coherence of idea systems other than our own" (Kuhn 1976/1977a, 11).

Consequently, I think that it is more likely that it was Kuhn's encounter with Aristotle's work that drew his attention to scientific revolutions and the complications they raise for a cumulative account of scientific progress.

My aim in this chapter has been to understand what Kuhn learned from the Aristotle experience he had in the summer of 1947, when he was preparing a set of lectures on the history of early modern mechanics as part of his contribution to the General Education Natural Science course that he was working on with Conant. Though the experience left Kuhn with an acute awareness of the existence of scientific revolutions, having had to cross a revolutionary divide to make sense of Aristotle's physics, he still lacked a clear sense of what role these episodes played in the *development* of scientific knowledge. That would come later, but only after he had developed an understanding of normal science and its relationship to revolutionary changes of theory. So, on the one hand, the importance of the experience should not be underestimated. It was the catalyst for writing *Structure*. It showed him that many of the assumptions he had held about science were mistaken; most importantly, the assumption that the growth of scientific knowledge is strictly cumulative. On the other hand, in a certain sense, the Aristotle experience was just a catalyst, for it would take Kuhn another fifteen years before he could articulate the significance of scientific revolutions in the development of scientific knowledge. Much more work would be required before he could do this.

Finally, as noted above, Kuhn also drew some methodological lessons from the Aristotle experience. These pertain to the practices of conducting research in the history of science. These lessons were, no doubt, important to Kuhn as a practicing historian, and in the seminar room, teaching the history of science. But they are quite tangential to the philosophical view that he was developing, initiated by the Aristotle experience. Historians of science are far less concerned with scientific revolutions than are philosophers of science (see, for example, Shapin 1996). They are, though, concerned with gaining the skills to cross a cultural (perhaps revolutionary) divide in their efforts to make sense of the work of earlier scientists and natural philosophers. Not surprisingly, given the profound impact that reading Aristotle had on his own development as a historian, Kuhn would often have students read Aristotle in his courses (see Kuhn 1997/2000, 288).

CHAPTER 2

The Influence of James B. Conant

My aim in this chapter is to examine the influence that James B. Conant had on Kuhn and on the views expressed in *The Structure of Scientific Revolutions*. To date, no one has carried out a systematic study of Conant's influence on Kuhn.[1] By clarifying Conant's influence on Kuhn, I also hope to clarify the influence that others had on Kuhn's thinking. A number of philosophers have suggested that Kuhn's principal contributions were anticipated by someone else. Ludwik Fleck, Stephen Toulmin and Michael Polanyi are sometimes named. I aim to set the record straight on this issue. Further, by identifying the various ways in which Conant influenced Kuhn's view of science we will be in a better position to understand Kuhn's most original contributions in *Structure*.

Surprisingly, Kuhn says very little about Conant's influence on his thinking. He does suggest that Conant was the smartest person he ever met (Kuhn 1997/2000, 259–260). But he never discusses how Conant influenced him.

On the one hand, I aim to show that much of the framework and many of the concepts that figure in *Structure* were part of Conant's picture of science, a picture that figured prominently in the General Education Natural Science courses that Conant taught with the assistance of others, including Kuhn when the latter was a graduate student in physics at Harvard. Consequently, I argue that it is likely that Conant is the principal source for many of Kuhn's ideas, rather than the various other sources commonly cited. On the other hand, I aim to show that Kuhn's *Structure* contains important contributions that do not figure in Conant's picture of science. I argue that the following five contributions in *Structure* were developed by Kuhn independently of Conant:

---

[1] Joel Isaac (2012) looks briefly at Conant and his influence on Kuhn, but Isaac's principal concern is with understanding the development of the social sciences at Harvard. Also, Justin Biddle (2011) provides a thorough study of Conant's "theory of science," but Biddle is not concerned with Conant's influence on Kuhn.

1) the concept of a paradigm;
2) the concept of normal science;
3) *the problem* of scientific revolutions, that is, the apparent threat posed by radical changes of theory in science;
4) the related concept of incommensurability; and
5) Kuhn's emphasis on the social dimensions of science, a focus that is largely responsible for the development of both the Strong Programme in the Sociology of Scientific Knowledge (SSK), and more recently the social epistemology of scientific knowledge.

Some Background

In an effort to better understand the influence that Conant had on Kuhn when he wrote *Structure*, it is worth first considering two popular claims about what influenced Kuhn when he wrote *Structure*. The first acknowledges the influence of Conant but focuses on Cold War culture. The second suggests that Kuhn's work is derivative of someone else's work. Both of these claims, I argue, are misguided.

First, Steve Fuller has famously argued that Kuhn's *Structure* can and should be read as a rationalization for a new social order in science, a social order tied to the Cold War (Fuller 2000, 5 and 70; see also Philip Mirowski 2005, 793). During the Second World War scientists were enlisted in great numbers to help with the war effort. In America this culminated in the Manhattan Project, the design, construction and employment of an atomic bomb. After the war it seemed that there would be no turning back. The government was now deeply dependent on science and scientists, and scientists had become accustomed to the greater resources and higher social status that came with working closely with and for the government. Indeed, the new postwar "superpower" status of the United States depended on the nation being a world leader in science.

Fuller claims that Kuhn's *Structure*, with its emphasis on scientific research communities as closed communities accountable only to themselves, created a picture of science that helped shield it from public scrutiny. The underlying message was that the public was unfit to evaluate science and scientists, despite the fact that the public's fate was in the hands of those scientists, and despite the fact that the public pays for scientific research. On this view, *Structure* is just a piece of propaganda, promoting a particular vision of science that Conant was instrumental in designing and implementing.

In fact, as Fuller notes, Conant was deeply involved in science policy in America during the war. Working with Vannevar Bush, the head of the Carnegie Institution in Washington, Conant was instrumental in overseeing "the growth of Big Science during [the Second World War]" (Isaac 2012, 204; see also Conant 1970, part III). Together, leading up to and during the Second World War, they served on the National Defense Research Committee, which was designed to "mobilize American scientific contributions to military research" (Hershberg 1993, 127; see also Biddle 2011, section 2.1). Conant thus helped *create* the new social order that depended on scientists working with and for the government. In the Terry lectures, Conant describes his involvement in wartime science as follows:

> as I watched the secret development of the atomic bomb through four years of war, I thought of the work being done at the same time under the auspices of the Medical Research Committee. I knew of the then secret research on penicillin, on DDT, on anti-malarial drugs, on the use of blood plasma, and realized how much these scientific advances meant for the future of mankind. (Conant 1946/1947, xi–xii)

So Fuller's suggestion is not wholly baseless.

However, Fuller fails to make clear the exact nature of the connection between Conant's wartime experiences and the views Kuhn presents in *Structure*. Indeed, at times, Fuller's reading of Kuhn borders on conspiracy theory. For example, Fuller claims that Kuhn's fusion of "the industrial and communal images of science . . . may have played a strategic role in the understanding of science that Conant wanted America's future leaders to have" (Fuller 2000, 213). This, and other aspects of Fuller's reading of Kuhn have met with some fierce criticism (see, for example, Andersen 2001).

Oddly, there is also an irony to Fuller's reading of *Structure*. As Fuller recognizes, Kuhn was insistent that in certain respects science is very conservative. That is a central theme in Kuhn's discussion of the essential tension. Science, he insisted, is tradition-bound, like all other social institutions. What intrigued Kuhn and what he sought to explain was how such a conservative institution was capable of innovation.

But this message was lost on the student radicals of the 1960s (see Fuller 2000, 74). They found Kuhn's book exciting, even liberating. The students were moved by the talk of revolutionary change and the possibility of freeing themselves from the grip of the paradigms in which they were trapped. These concerns were no part of Kuhn's concerns when he wrote *Structure*. Kuhn was quite insistent. The book was about the natural sciences, narrowly and exclusively. He did not even think his theory of science described the social sciences, let alone society at large. Hence, if Fuller's reading is correct, the public, at least the radical public, missed the message. They did not see the book as supporting the status quo. Indeed, Kuhn certainly thought the radical students missed his point.

There is an additional problem with Fuller's reading of *Structure*, specifically the book's alleged role as propaganda for the new order of science. According to Fuller, Kuhn's account of science provides a rationale for isolating and protecting science and scientists from public scrutiny. Fuller suggests that this was necessary in order to ensure that the public did not object to the employment of scientists by the state; in short, the co-opting of science by the state. Fuller thus regards Conant's influence on Kuhn as principally, if not exclusively, ideological. But, contrary to Fuller's reading, there is evidence that Conant's commitment to protecting the autonomy of science and scientists was motivated by a very different concern. According to James Hershberg, Conant was concerned with protecting scientists, and academics in general, from interference and persecution during the McCarthy era anti-communist witch hunts (Hershberg 1993, 406). Conant believed that scientists cannot do their research effectively if they are in constant fear of being branded communists or communist sympathizers. They need the freedom to pursue their research wherever it may take them, unencumbered by the fear of persecution.

In fact, even before the end of the Second World War, Conant had already voiced concerns about allowing the government too much power in directing science. Daniel Kevles cites a speech Conant gave in 1943, titled "Science and Society in the Post-War World": "Beware in times of peace ... of coordinating agencies with dictatorial powers – of ideas of a peacetime scientific general staff" (Conant 1943, cited in Kevles 1977, 12). Indeed, Vannevar Bush voiced similar sentiments in his book *Endless Horizons*, published in the wake of the Second World War. Bush had played an integral role in getting America ready for the war, and he was insistent that in times of peace the government should not direct scientific research (see Bush 1946/1975, especially chapter 11).

There is an alternative theme in writings about Kuhn's influences. It is not uncommon for critics and commentators to argue that Kuhn was anticipated by someone else, usually someone who has not received the recognition they deserve. These studies are motivated by the belief that the originality of *Structure* has been greatly exaggerated, and certainly exaggerated beyond its influence. Fleck (1935/1979), for example, is often cited as having anticipated Kuhn, though more careful studies suggest that the similarities between their views are superficial (Brorson and Andersen 2001; Babich 2003; and Mößner 2011). What is interesting about the case of Fleck is that it is Kuhn who is largely responsible for making Fleck's work known. Kuhn refers to Fleck's *Genesis and Development of a Scientific Fact* in the Preface to *Structure*. As noted in Chapter 1, Kuhn acknowledges that this work was instrumental in helping him see that his own project was tied to "the sociology of the scientific community" (1962/2012, xli). Indeed, Kuhn even acknowledges that Fleck anticipated "many of [his] own ideas" and that he was unable to "reconstruct or evaluate" the full influence that Fleck's book may have had on his own thought (1962/2012, xli). Further, Kuhn acknowledges that he "certainly got a lot of reinforcement" from reading Fleck's book (see Kuhn 1997/2000, 283). Elaborating, Kuhn notes that "it was very important that I read that book because it made me feel, all right, I'm not the only one who's seeing things this way" (Kuhn 1997/ 2000, 283).

Thus, Kuhn in no way attempted to hide the fact that he was influenced by Fleck. Kuhn was also instrumental in seeing that Fleck's book was translated into English, and thus reached a wider readership. In fact, until the English translation was available, it was both largely neglected and difficult to find. Thaddeus Trenn, one of its translators, notes "Fleck received no notices at all in a sampling of standard works in the history of medicine and the sociology of knowledge. Indeed, the first published reference to his monograph since the 1930s seems to be that of Kuhn in 1962" (Trenn 1979, xviii). And in 1967, David Edge, then teaching in the Science Studies Unit at the University of Edinburgh, wrote to Kuhn about Fleck's book. Anticipating that Kuhn might know whether an English translation was in the works, Edge noted that "no library in Edinburgh has a copy" of the German edition (Edge 1967). In fact, even Robert K. Merton, who read prodigiously, claims that he had never heard of *Genesis and Development of a Scientific Fact* until Kuhn mentioned it (see Merton 1976). Constructing counterfactual histories is notoriously problematic, but it is hard to imagine that Fleck's book would have had much impact if Kuhn had not drawn attention to it.

Similarly, some have argued that Kuhn borrowed generously from Toulmin and Polanyi. Polanyi raised this concern in two personal letters, one to Donald Campbell and the other to Paul Halmos, intimating that he believed that Kuhn may not have given him adequate credit given the extent to which *Structure* draws on concepts from Polanyi's earlier writings (see Timmins 2013, 311; see also Nye 2011, 243). A number of the articles that aim to show Kuhn's debt to Polanyi are more inflammatory, suggesting that Kuhn knowingly borrowed from Polanyi without giving Polanyi the credit he was due (see Jacobs 2007 and Moleski 2007). These studies thus aim to set the historical record straight. However, they suffer from two types of flaws. First, these studies embody a tension that undermines the credibility of the accusations against Kuhn. They claim that Kuhn borrowed from Polanyi, yet they also insist on key differences between the views of Kuhn and Polanyi. If their views are as different as these studies suggest, then it is difficult to understand how one can be indebted to the other. Second, these studies fail to take into account other influences on Kuhn's thinking, for example, Conant's influence.

Incidentally, Marjorie Grene did not think that Kuhn plagiarized Polanyi. This is significant, as she knew Polanyi's work intimately, and was a friend of his (see Mullins 2009/2010; 2002). In fact, Grene was warmly thanked by Polanyi in the acknowledgments of his book *Personal Knowledge* (see Polanyi 1962, ix). Anyone who had met Grene would have known that she would not have hesitated to call someone out on plagiarism or any other sort of misconduct if there were suspicions of it (see Mullins 2009/2010, 35). Instead, in a letter to Kuhn dated September 25, 1963, Grene notes that "having finally acquired your *Structure of Scientific Revolutions*, I can't resist just writing to tell you how delightful I find it" (Grene 1963). If there were a case to argue on Polanyi's behalf, Grene would almost certainly have taken it on. Incidentally, in this same letter Grene also expresses surprise at where the book is published. "It seems a bit of a joke that it should appear in the Unity of Science Series, of all places. But I gather that even the *Minnesota Studies* are gradually coming our way" (Grene 1963).[2]

The Toulmin case is also perplexing, as Toulmin does not appear to accuse Kuhn of appropriating his ideas without giving due credit. And in two publications in which he discusses Kuhn's *Structure*, Toulmin emphasizes the differences between their views. In his contribution to the 1965

---

[2] I think Kuhn is somewhat responsible for some of the confusion in the case of Polanyi. On the one hand, at the end of his life, Kuhn remarked that as he was writing *Structure* and came across Polanyi's *Personal Knowledge*, he "looked at it and said, I must not read this book now. I would have to go back to first principles and I wasn't going to do that" (Kuhn 1997/2000, 296–297). On the other hand, Kuhn does in fact cite Polanyi's book in *Structure*. Kuhn claims that "the existence of a paradigm need not even imply that any full set of rules exists" (Kuhn 1962/2012, 44). He then explains in a footnote that "Michael Polanyi has developed a very similar theme, arguing that much of the scientist's success depends on 'tacit knowledge,' i.e., upon knowledge that is acquired through practice and that cannot be articulated explicitly" (Kuhn 1962/2012, 44–45, fn. 1). So it appears that Kuhn did in fact read the book before completing *Structure*.

Bedford College conference volume, Toulmin presents a view of scientific change modeled on an *evolutionary* process, contrasting it with Kuhn's account, which emphasizes the role of *revolutionary* changes (see Toulmin 1970/1972). Toulmin thus saw their views as fundamentally different. It is debatable whether it is correct to contrast evolutionary and revolutionary accounts of scientific change, as if they were mutually exclusive. And I have argued in detail that Kuhn's account has evolutionary and revolutionary dimensions (see Wray 2011, chapter 8). The important point for our purposes is that Toulmin regarded Kuhn's account as different from his own, and thus did not think that Kuhn had borrowed from him in some sort of questionable manner.

And in an extended discussion of the shortcomings of Kuhn's view in *Human Understanding*, Toulmin compares Kuhn's view to Collingwood's, raising objections to both (Toulmin 1972, 98–123). Again, it is difficult to reconcile his critique of Kuhn's view with an accusation of Kuhn appropriating Toulmin's views. In the mid-1990s, in an autobiographical interview, Kuhn acknowledges that he can understand why Toulmin might have been led to believe that he had stolen some of his ideas, but Kuhn both denies that he did steal from Toulmin, and notes that Toulmin never made such an accusation (see Kuhn 1997/2000, 296–297).

Further, Kuhn readily admits that he was influenced by Fleck, Toulmin and Polanyi (see Kuhn 1997/2000, 296–297). However, one of my main goals in this chapter is to show that many of the concepts and themes in Kuhn's *Structure* that one might be tempted to trace to either Fleck, Polanyi or Toulmin probably have their roots in Conant's writings.

Conant's View of Science

In this section I want to examine some of Conant's writings, writings that Kuhn would have been very familiar with, in order to identify themes and views that shaped Kuhn's own thinking about science. Conant was a scientist, specifically a chemist, and thus familiar with the work of scientists in a very intimate way. He was the president of Harvard, and he was also a leader in science policy at a time when the United States government was looking more and more toward scientists for policy advice. Not surprisingly, Kuhn had a great deal of respect for Conant (see Kuhn 1997/2000, 259–260; 274; 275).

As mentioned in Chapter 1, Kuhn worked closely with Conant on the General Education science courses at Harvard. From 1947, when Kuhn was still a graduate student in physics, he was responsible for "a set of lectures on the origins of seventeenth-century mechanics" (see Kuhn 1977, xi). It was as Kuhn prepared for his teaching responsibilities in this course that he had his famous Aristotle epiphany, when he realized that Aristotle was a very different sort of physicist than Newton was (see Kuhn 1977, xi–xii). Kuhn came to realize that one needed to understand a theoretical framework from within, a task that can be more challenging than one may think, especially when the framework uses terms that one is already familiar with, but with different connotations (and sometimes even different denotations). In the case of Aristotle, Kuhn had to realize that "the subject that was to become mechanics was ... a special case" of a broader subject that concerned Aristotle in *Physics*, "change-of-quality in general," which "included both the fall of a stone and the growth of a child into adulthood" (Kuhn 1977, xi). In order to understand Aristotle's physics one must first realize that Aristotle was not aiming to develop Newtonian mechanics. His project was fundamentally different.

The General Education Natural Science course was an initiative that Conant took very seriously (see Conant 1970, 367–373). He felt that a General Education course in the sciences would be the means by which tomorrow's leaders, that is, the Harvard students in his classroom, would be able to conduct their jobs effectively, whether in business or in politics (Conant 1947, 1; Isaac 2012, 205; and Biddle 2011, 554). Conant explicitly mentions college "graduates who are to be lawyers, writers, teachers, politicians, public servants, and businessmen" as the target audience for the course (Conant 1947, 1). Science was now playing such an important role in the lives of American citizens that it was essential that leaders had a sense of what it was and how it worked. The General Education science courses were designed with that purpose in mind. Conant proposed using a proven method of pedagogy, "the case method," a method used to great effect in law schools around the nation and at the Harvard Business School (Conant 1947, 17; see also Isaac 2012, 75–80).

John Shirley (1951) provides an insightful contemporary assessment of the *Harvard Case Histories in Experimental Science*, including the development of the General Education science courses at Harvard during the mid-1940s and early 1950s. On the one hand, Shirley notes "without question, President Conant is ... proceeding in a scientific manner to handle philosophic ideas for the nonscientific, nonphilosophic audience. The experiment is a noble one; it deserves and will have success" (Shirley 1951, 422). On the other hand, Shirley raises the concern that limiting "the materials ... to published works ... excludes two very important sources of information about [science]: manuscripts and machines. Science is not now, nor has ever been, the exclusive prerogative of the theorist; yet it is the scientific philosopher who has been the publisher of scientific treatises" (Shirley 1951, 423).

Conant's efforts in developing the General Education Natural Science course can be seen as part of a contemporary concern with the widening gap between the "two cultures," the humanities and the natural sciences. Long before C. P. Snow delivered his influential lectures, the historian of science, George Sarton, expressed similar concerns to Conant. In a 1943 letter to Conant, Sarton claimed that "the widening abyss separating [humanists and scientists] must be bridged or our culture will be jeopardized and defeat itself" (cited in Hershberg 1993, 409).

The aspects of Conant's view of science relevant to my analysis here were developed and articulated in a series of publications between 1947 and 1953. In his Terry lectures at Yale University, subsequently published as *On Understanding Science: An Historical Approach*, Conant outlines how one might teach a course in the history of science with the aim of teaching non-science majors the "tactics and strategies of science." In fact, Conant was using *On Understanding Science* in the General Education Natural Science course when Kuhn began to work with him (see Kuhn 1997/2000, 275). Conant elaborates on his view of science in the "Forward" to the *Harvard Case Histories in Experimental Science*, the text used in the General Education science courses at Harvard, a document with which Kuhn would have been very familiar (see Conant 1950). Further refinements were made in the early 1950s when Conant was invited to give the Bampton lectures at Columbia University. These lectures were intended to engage an educated general audience, and were published as *Modern Science and Modern Man* (Conant 1953). There are eight themes and views in these publications that were either adopted by or coincided with Kuhn's own views as published in *Structure*.

Theme (1)   The Historical Approach

Conant had a significant influence on Kuhn's methodology. Conant saw great value in studying the history of science as a means to understanding how science works (Conant 1953, 53). He contrasts the historical approach to understanding science, which aims "to retrace the steps by which certain end results have been produced," with a logical or philosophical approach, which aims "to dissect the results with the hope of revealing its structural pattern and exposing the logical relations of the component parts" (Conant 1947, 12).

The former method, he suggests, is generally the more fruitful one (Conant 1947, 12). Indeed, Conant believes that the latter approach to studying science can be misleading (Conant 1947, 13–14). Given the importance of positivism at the time Conant was writing, it is worth quoting Conant at length.

> I doubt if the philosophical treatments of science and scientific method have been very successful when viewed as an educational enterprise. No one questions ... the importance of this type of penetrating analysis. There must be constant critical appraisal of the progress of science and in particular of scientific concepts and operation. This is one of the prime tasks of philosophers concerned with the unity of science ... But when learned discussions of these difficult matters are the sole source of popular knowledge about the ways of science, education in science may be more handicapped than helped by their wide circulation. (Conant 1947, 13–14)

Conant thought that "the historical approach" to the study of science was the key to appreciating the "dynamic quality of science" which, at its core, involves the "development of conceptual schemes" (Conant 1947, 24).

Kuhn was profoundly influenced by Conant in this regard. Indeed, Kuhn's involvement with Conant in teaching the General Education Natural Science course at Harvard was crucial to Kuhn's decision to change careers and become a historian of science. And Kuhn's approach to the study of science was and always remained deeply historical (see Hoyningen-Huene 1992). Throughout *Structure*, Kuhn emphasizes the new insights we can gain into science and scientific knowledge if we view it from a historical perspective. Most importantly, the history of science can shift our focus away from "finished scientific achievements," the traditional focus of those who study science (Kuhn 1962/2012, 1).

In fact, most of the sources cited in *Structure* are sources in the history of science (see Wray 2015c). To be precise, such sources account for 60 percent of the citations. The standard origins story of the historical school in philosophy of science traces its initial development to Hanson, Toulmin, Feyerabend and Kuhn, who reacted against the logical analyses of science that were the focus of the Logical Positivists and Logical Empiricists. Indeed, Kuhn repeats this story himself on numerous occasions. We can see now that at least one strand of this traditional origins story can be traced further back to Conant. Indeed, as we saw above, already in the 1940s, Conant was drawing attention to the limitations of the Logical Positivists' analyses of science.

Theme (2)   Conceptual Schemes

Conant conceived of scientists as working with conceptual schemes (Conant 1947, 19, 24). Conceptual schemes or frameworks aid scientists in organizing their experiences and observations. As Conant expresses the point, "science is a process of fabricating a web of interconnected concepts and conceptual schemes arising from experiment and observation and fruitful of further experiments and observations" (Conant 1953, 106–107; see also Conant 1950/1965, 4). Conant's choice of the term "conceptual scheme" is very deliberate. He notes

> the words *theory* and *hypothesis* are frequently employed to describe both conceptual schemes and models, or pictures, "explaining" such schemes. A discussion of the definition of theory and hypothesis is often given in elementary texts, but I think such considerations of doubtful value. Because of the ambiguity it is well ... to avoid as far as possible both words in discussing the evolution of science. (Conant 1947, 47; emphasis in original)

Conant believed that scientists devote significant efforts to reconciling the facts with the accepted conceptual scheme. In his analysis of the chemical revolution that was ushered in by Lavoisier, Conant notes how the early chemists sought to fit the facts to the then-accepted phlogiston theory. As Conant explains, the chemists working with the conceptual scheme associated with the phlogiston theory "were interested in reconciling one inconvenient set of facts with what seemed from their point of view an otherwise admirable conceptual scheme" (Conant 1947, 89).

Though Kuhn used Conant's term "conceptual scheme" in his first book, *The Copernican Revolution*, he does not use the term at all in *Structure*. But, like Conant's, Kuhn's analysis of science focuses on theories or theoretical frameworks as the locus of change in scientific research communities. When a research community is conducting normal scientific research, a particular theory holds the allegiance of the community. And revolutionary changes of theory are precipitated by crisis, when the accepted theory no longer provides unequivocal guidance to those working in the research community (on normal science, see Kuhn 1962/2012, chapters 1–4; on crisis, see Kuhn 1962/2012, chapters 7–8). And like Conant, Kuhn claims that scientists spend much of their careers making nature fit into the conceptual boxes supplied by the accepted theory (see Kuhn 1962/2012, chapter 3, especially p. 24). Paul Hoyningen-Huene has suggested that Kuhn may be indebted to Conant for this focus on theoretical frameworks or theories (1989/1993, 199–200, n. 13).

Theme (3) Holism

Conant was a holist of sorts. Specifically, he did not believe that scientific facts speak for themselves (Conant 1953, 113). He also appreciated the fact that tests of hypotheses are mediated by background assumptions that are taken for granted. Conant illustrates the complicated way in which theories are tested with an analysis of Pascal's test of the hypothesis that "the earth is surrounded by a sea of air which exerts pressure" (1953, 121–122). Such a hypothesis, Conant notes, "was not and could not be tested directly" (1953, 122). Instead, "the limited working hypothesis that was tested was, essentially, 'If I set this barometer here on the summit [of Puy de Dôme] and measure the height of the mercury column, it will be less than that observed at the base'" (Conant 1953, 122). Conant notes "the connection between the verification or negation of this limited hypothesis and the broad working hypothesis . . . involves many steps and assumptions" (1953, 122–123).[3]

Similarly, Kuhn repeatedly emphasized the theory-laden nature of observation, the fact that what scientists see is to a great extent determined and limited by the theoretical presuppositions they bring to their inquiries. Chemistry, Kuhn suggests, was able to make significant progress after chemists began imposing the law of fixed proportions onto the world (1962/2012, 132–134).

In all fairness, there is reason to believe that others besides Conant influenced Kuhn with respect to holism. One source is probably Max Weber. Kuhn read Weber's *Methodology of the Social Sciences* in 1949. In a notebook recording his readings and thoughts during this period, Kuhn describes Weber's book as "continually brilliant," citing as an example Weber's remark that "'the establishment of such regularities [i.e. scientific laws] is not the *end* but rather the *means* of knowledge'" (Kuhn 1949). Thus, Kuhn believed that theories help scientists find order in the world.[4]

---

[3] Justin Biddle suggests that Quine is probably the key source of Conant's holism (see Biddle 2011, 556). I have reservations about this. First, as Biddle notes, "the extent to which Conant was familiar with [Quine's holism] is not completely clear" (see Biddle 2011, 556). Second, the various citations to Quine that Biddle draws attention to in Conant's writing are all to Conant's *Scientific Principles and Moral Conduct*, which was published in 1967 (see Biddle 2011, 556 and 557). Conant's holism, though, can be traced back to the early 1950s.

[4] Bojana Mladenović suggests that Kuhn also drew on Weber's theory of ideal types (see Mladenović 2017).

Theme (4) Scientific Revolutions

Conant believed that scientific revolutions were a major source of the greatest advances in science. In Conant's words, "the history of science demonstrates beyond a doubt that the really revolutionary and significant advances come not from empiricism but from new theories" (Conant 1953, 53). Importantly, though, Conant believed that experimental work is often the impetus for the development of new concepts, and thus new conceptual schemes (see Conant 1947, 59). For example, Conant argues that Boyle's experimental work led to the development of "the idea of an atmosphere exerting pressure, and the concept of air as an elastic fluid," concepts that replaced the "Aristotelian *horror vacui*" (Conant 1947, 59). Conant generalizes from this and similar cases, arguing that "tremendous spurts in the progress of the various sciences are almost always connected with the development of a new technique or the sudden emergence of a new concept" (Conant 1947, 73–74). Conant also notes that sometimes an accepted conceptual scheme can have "an almost paralyzing hold on [scientists'] minds," thus making a change of conceptual scheme in a research community particularly challenging (1947, 80; see also 97 and 103).

Kuhn also believes that significant advances in science result from revolutionary changes of theory. In comparison to Conant, though, Kuhn was less insistent that revolutions are the sources of the greatest progress. Indeed, he was less sanguine than Conant about narratives of progress in science, especially with respect to theoretical knowledge. A principal concern of Kuhn's was to undermine the popular notion that scientific progress is a strictly cumulative process, always involving the addition of new knowledge to old. For example, Kuhn insisted that the revolutionary changes of theory in the history of physics did not constitute a series of steps in a march ever closer to the truth. In the Postscript to *Structure* Kuhn is explicit that, though he does "not doubt ... that Newton's mechanics improves on Aristotle's and that Einstein's improves on Newton's as instruments for puzzle-solving ... in some important respects ... Einstein's general theory of relativity is closer to Aristotle's than either of them is to Newton's" (Kuhn 1969/2012, 205). Kuhn would instead characterize changes of theory as profoundly redirecting the course of research in a field, and in the process, resolving some persistent anomalies.

What is more, like Conant, Kuhn also believed that theories could have a strong grip on scientists' minds. Indeed, this is what Kuhn's appeal to the anomalous playing-card experiments conducted by Bruner and Postman is intended to illustrate. Some of the experimental subjects had great difficulties identifying the anomalous playing cards as anomalies. Scientists, according to Kuhn, are equally prone to the grip of the preconceptions they have internalized in learning the theory with which they work.

Theme (5)   Unified Science

Conant was skeptical about the possibility of a unified science (1953, 75). In a brief analysis of early twentieth-century studies of the atom, Conant argues that "it is conceivable that we might have had to be operating with two sets of models, one to assist the chemist in formulating the results of his labors, the other to guide the investigations of the physicists" (Conant 1953, 75). Generalizing, Conant did not believe that it is a foregone conclusion that scientists in two neighboring disciplines will make the same assumptions about the properties of the entities they study. And, importantly, Conant does not present this as a failing on the part of scientists. Rather, he wants us to see that unified theories are not inevitable outcomes of advances in science.

   Kuhn was even more reticent to assume a unified science. For Kuhn, the unit of analysis is the scientific community, that is, the specialty. These are the groups of scientists that work with the same theory. It is the dynamics of their research that he is tracing in his cyclical pattern from normal science to crisis, from crisis to revolution, and from revolution to a new normal scientific tradition. There is no presumption that the theories of all specialties can be rendered or articulated in the same vocabulary or set of concepts. Indeed, in his later writings this comes out more explicitly. Kuhn talks about the incommensurability between the theories in neighboring specialties (see Kuhn 1991/2000a, 98; see Wray 2011, 74–76, for a fuller discussion of this topic). He came to believe that this sort of incommensurability served an important function, allowing the scientists in a new scientific specialty to develop conceptual tools that serve their research interests.

Theme (6)   Fruitfulness

Conant emphasized the importance of fruitfulness in the evaluation of theories. He characterizes a theory's fruitfulness as its ability "to suggest, stimulate, and direct experiment" (Conant 1953, 92; see also Conant 1950/ 1965, 4). Conant again emphasizes the importance of fruitfulness in a 1961 letter addressed to Kuhn regarding the draft manuscript of *Structure*. Conant remarks that when there is a change of theory in a field, "the new theory is accepted not only because the deductions from it fit more 'facts' . . . than the older theory but because working with it . . . opens more new vistas" (Conant 1961).

The importance of the theoretical virtue of fruitfulness is repeatedly emphasized by Kuhn. Though he does not use the term in this passage from the Postscript, it is clear Kuhn is referring to fruitfulness when he claims that one of the virtues that scientists look for in theories is that "they permit puzzle-formation and solution" (Kuhn 1969/2012, 184). The notion of fruitfulness becomes even more explicit in Kuhn's later writing, especially, for example, "Objectivity, Value Judgment, and Theory Choice," in which fruitfulness is presented as one of the five canonical theoretical values, values that scientists have accepted throughout the history of science and across the disciplines (Kuhn 1973/1977, 321–322).

Theme (7) Scientific Method

Conant is critical of the notion of "*the* Scientific Method" (1947, 14; 1953, 39). Instead, he speaks of "the *methods* by which science has been advanced" (Conant 1947, 20; emphasis added). Conant objected to the common view that there is a single method guiding all scientists that can be followed almost mechanically. He also thought the scientific method, as typically described, exaggerated the role of observation and experience in science. Contrary to the impression one gets from mechanical accounts of the scientific method, Conant argues that "even the ablest of the early scientists had to fight through the thickets of erroneous observations, misleading generalizations, inadequate formulations, and unconscious prejudice" (1947, 15). Moreover, he believed that advances in science, and modern science in particular, were made by moving beyond observations and experiences. He repeatedly claims that "as . . . sciences became more equipped with more and more satisfactory theories, the degree of empiricism . . . diminished" (1953, 48).[5]

Like Conant, Kuhn is also skeptical about the role of the so-called "scientific method" in research. Kuhn does argue, after all, for the priority of paradigms. In this context, "paradigm" means exemplar. Kuhn believes

---

[5] Justin Biddle rightly notes that there are two distinct claims that Conant was attacking with his attack on *the* scientific method: "(1) there is a single procedure, or set of rules, that scientists follow (or should follow), and (2) following this procedure will ensure that inquiry is dispassionate and unprejudiced" (Biddle 2011, 555).

that it is past scientific achievements that guide scientists in their pursuit of a solution to a research problem, not some method that can be followed mechanically or expressed in an algorithm (see Kuhn 1962/2012, chapter 5). Reasoning from paradigms is not algorithmic. Instead, it involves analogical reasoning, the sort of reasoning one engages in when one appeals to metaphors (see Hacking 2012, xix; also Hacking 2016). Kuhn's discussion of the priority of paradigms in *Structure* is thus meant to shift emphasis away from scientific method. Indeed, Biddle has already suggested that Conant anticipated Kuhn in attacking the notion of *the* scientific method by appealing to the history of science (see Biddle 2011, 553).

Theme (8)   Theory Evaluation Is Comparative

Conant believed that "a theory is only overthrown by a better theory, never merely by contradictory facts" (Conant 1947, 36). Conant thus acknowledges that an accepted theory may not be able to account for all the relevant facts, and that this is not sufficient grounds for rejecting it. He claims that "a conceptual scheme is never discarded merely because of a few stubborn facts with which it cannot be reconciled" (Conant 1947, 84). Moreover, he claims that "it takes a new conceptual scheme to cause the abandonment of an old one: when only a few facts appear to be irreconcilable with a well established conceptual scheme, the first attempt is not to discard the scheme but to find some way out of the difficulty and keep it" (1947, 89; see also 103). And he reminds us that "sometimes it turns out that difficulties with a concept or conceptual scheme are wisely ignored" (Conant 1947, 96). Sometimes a fact that was initially recalcitrant can be accommodated by the accepted conceptual scheme once appropriate adjustments have been made to that scheme. Because of the important role conceptual schemes play in science and the growth of scientific knowledge, guiding scientists in their research, scientists will only reject a theory when they have a new conceptual scheme ready to take its place. Conant explains that "only the combination of a new concept with facts contradictory to the old ideas finally brings about a scientific revolution" (1947, 36–37; see also 84).

Again, Kuhn frequently emphasized the fact that theory evaluation is comparative. "The decision to reject one paradigm is always simultaneously the decision to accept another, and the judgment leading to that decision involves the comparison of both paradigms with nature *and* with each other" (Kuhn 1962/2012, 78). Hoyningen-Huene suggests that this belief, that scientists are reluctant to discard a theory until they have a new one to take its place, originated with Conant (Hoyningen-Huene 1989/ 1993, 237, n. 82).

Thus, we can see the many similarities between Conant's and Kuhn's thinking about science. These similarities include both matters of methodology in the study of science, and the content of their thoughts about scientific inquiry. Given Conant's seniority, it is likely that the influence ran from Conant to Kuhn. And given that Kuhn worked with Conant in 1947, it is likely that the principal influence on Kuhn was Conant, not Fleck, Toulmin or Polanyi.

Before considering Kuhn's own contributions, I want to briefly discuss Conant's influence on Kuhn's first book, *The Copernican Revolution*. This book was developed in part as a contribution to Conant's General Education Natural Science course, which as we have seen, was a course designed for non-science students (see Kuhn 1956/1957, viii–ix).

In addition to Kuhn's responsibility for lectures on the origins of seventeenth-century mechanics, by 1949 he was also responsible for lectures on the Copernican Revolution in astronomy in the General Education Natural Science course. Because the book was more or less written in the context of that course, it is not surprising that *The Copernican Revolution* uses much of the same vocabulary that Conant uses. But, it is noteworthy that a number of the key concepts that would figure in Kuhn's analysis in *Structure of Scientific Revolutions* do not appear in *The Copernican Revolution*, for example, "paradigm," "normal science" and "incommensurability" (see also Melogno and Courtoisie 2019, 27). Recall, from Chapter 1, that Kuhn did not even develop the notions of normal science and paradigm until after *The Copernican Revolution* was published.

In the earlier book Kuhn even adopts Conant's expression "conceptual scheme" to describe the theoretical framework that scientists work with (Kuhn 1957, 3). Like Conant, Kuhn sees the replacement of one conceptual scheme or theory with another as playing an integral part in the advancement of science (Kuhn 1957, 3). Kuhn also embraces Conant's holism. Kuhn claims "observation is never absolutely incompatible with a conceptual scheme" (Kuhn 1957, 76).

And already in *The Copernican Revolution* Kuhn had assimilated Conant's notion of "fruitfulness" (see Kuhn 1957, 40). Kuhn describes fruitfulness, that is, the ability of a theory to "guide a scientist into the unknown, telling him where to look and what he may expect to find," as "the single most important function of conceptual schemes in science" (see Kuhn 1957, 40). Importantly, he argued that "a scientist must believe in his [conceptual] system before he will trust it as a guide to fruitful investigations of the *unknown*" (Kuhn 1957, 75). Thus, when he wrote *The Copernican Revolution*, Kuhn believed that without the conviction that the theory one is working with reflects the structure of the world, a scientist has no reason to expect that they can derive any insight into the features of the world from the features of the theory.

But even here, Kuhn's interests departed from those of Conant. Kuhn was especially interested in studying the cognitive factors that affect the way scientists see the world and respond to discoveries (Kuhn 1957, 39). For example, Kuhn notes that conceptual schemes have a logical function and a "purely psychological function" (Kuhn 1957, 39). From a logical point of view, Kuhn claims that scientists seek economical conceptual schemes. From a psychological point of view, they seek "cosmological satisfaction" (Kuhn 1957, 39). A scientist wants to explain and understand the appearances.

Kuhn's interest in and engagement with the psychology of science would continue to grow as he discovered ways in which psychological studies offered insight into the grip that an accepted conceptual scheme has on perception, as well as the resistance and anxiety that scientists feel when the long-accepted conceptual scheme is discovered to be unfit to account for recalcitrant observations. In *Structure*, though, Kuhn would draw more heavily on research in psychology as a means to understanding how scientists' perceptions and thinking are shaped by their beliefs and scientific training.

The Copernican Revolution is clearly a product of Conant's course, but much of what is central to Kuhn's understanding of science in *Structure* is absent in the earlier book.

Kuhn's Contribution

After this long list of claims and views associated with Kuhn that may have originated with Conant, one may well wonder what was left for Kuhn to contribute to our understanding of science. Indeed, Kuhn is unequivocal that he had great respect for, in fact, admired Conant. Kuhn claimed: "I thought for many years that except for James Conant, [my father] was the brightest person I had ever known" (see Kuhn 1997/2000, 259–260). This assessment of Conant is not unique. For example, when the chemist Paul Bartlett was a graduate student at Harvard, he was equally impressed with Conant, claiming: "I came to think of Conant as probably the most truly intelligent man I ever knew" (Bartlett 1983, 106). And, when Kuhn recalled his experience with the General Education Natural Science course later in life, he remarked: "who the hell wouldn't have taken the chance to work with Conant for a semester?" (Kuhn 1997/2000, 274). So it is not a surprise that Conant would have had a profound effect on him and his thinking about science.

But there is still a lot that is original in Kuhn's account of science. Indeed, having isolated Conant's influences on Kuhn, I think we are now in a better position to identify Kuhn's most original contributions to the philosophy of science. I will focus on five key contributions.

One concept that clearly owes nothing to Conant is the "paradigm" concept. In fact, Conant was initially very uncomfortable with Kuhn's use of the concept when he read the draft manuscript of *Structure*. He thought the concept "paradigm" was obfuscating. In a letter to Kuhn written in 1961 Conant warns Kuhn that he would be remembered as "the man who grabbed on to the word 'paradigm' and used it as a magic verbal wand to explain everything!" (cited in Cedarbaum 1983, 173; see Conant 1961). In that letter, Conant encouraged Kuhn to restrict the use of the term "paradigm" to models and examples, and not apply it to books and theories (Conant 1961). Conant thus anticipated some of the problems that Margaret Masterman (1970/1972), Dudley Shapere (1964) and others would raise (see Wray 2011, 55–56). But Conant would later change his attitude toward Kuhn's use of the term "paradigm." Upon receiving a printed copy of *Structure*, and rereading it, Conant notes in a letter to Kuhn that "many of the points I found difficult a year ago [when I read the draft manuscript] now seem to me to present no difficulty" (Conant 1962). Conant continues, noting that

> I must admit frankly you choosing the word "paradigm" still troubles a little. Yet as I finish the book I begin to realize you perhaps had no other way to get your basic theme across ... I shall be curious to see how the philosophers and historians of science react to your choosing this word as well as their reactions to your general thesis. (Conant 1962)

Conant's misgivings were not ungrounded. Kuhn was forced to spend the next ten years clarifying what he meant by the term "paradigm." By the mid-1970s Kuhn had restricted the extension of the term "paradigm" to what he would from then on refer to as "exemplars," those concrete solutions to scientific problems that are used to model solutions to related research problems (see Wray 2011, chapter 3). During those ten years, Kuhn discovered what he meant by the term "paradigm," thus exemplifying his own theory of scientific discovery, according to which a discovery is a complex process extending over time.

A second important concept that originated with Kuhn is the concept of "normal science." The "paradigm" concept is so intimately tied to normal science that one could fail to appreciate the importance of the notion of normal science in its own right. Paradigms define normal scientific traditions, and it is their failure that leads to the breakdown of such traditions. Normal science is what young scientists-in-training are being trained for when they learn the accepted paradigms. And the typical scientist's career is spent addressing the problems of normal science (Kuhn 1962/2012, chapters 3–4). It is the breakdown of normal science, precipitated by crisis, which ultimately makes a revolutionary change of theory possible. And as far as Kuhn is concerned, the reason one might get the impression that the growth of scientific knowledge is cumulative, with no setbacks, is that one looks at normal science only. Again, there is reason to believe that Kuhn's use of this concept owes little to Conant. Indeed, we saw in the previous chapter that Kuhn only hit upon the notion of normal science in the late 1950s, after he had left Harvard for Berkeley.[6]

Paul Hoyningen-Huene suggests that, "in essence, *the juxtaposition* of normal and revolutionary development can already be found in Conant" (see Hoyningen-Huene 1989/1993, 167, fn. 1; emphasis added). Indeed, as Hoyningen-Huene suggests, some *aspects* of Kuhn's notion of normal science trace their origins to Conant (see, for example, Conant 1947, 36). For example, Kuhn would have been familiar with Conant's view that some developments in science are revolutionary and depend on radical conceptual innovations, and other developments can be readily accommodated by the accepted conceptual scheme. But this does not capture either the whole of Kuhn's concept of normal science, nor the essential features of it. For Kuhn, normal science is more than just those periods or phases in which there is a steady accumulation of scientific knowledge. The aspects of normal science that originated with Kuhn are: (i) the practice of solving research puzzles by reasoning from accepted paradigms, or exemplars, and (ii) the fact that the accepted theory is assumed to be more-or-less correct with respect to what it says about the world. These aspects of normal science are not captured by Conant's juxtaposition of normal and revolutionary science. Thus, I want to emphasize that the originality of the concept "normal science" should not be underestimated.

---

[6] Joel Isaac (2012) notes that the notion of "normal science" made its "debut" in Kuhn's paper "The Function of Measurement in Modern Science," a paper published the year before *Structure* was published.

Karl Popper explicitly notes Kuhn's originality here. Popper was not one to readily compliment his contemporaries. Indeed, he was known to be quite adversarial, challenging others' claims to originality (see Hacohen 2002, 209–210, especially notes 115 and 116; and Hacohen 2002, 211). But Popper admits to "feelings of indebtedness to Kuhn" for this discovery (see Popper 1970/1972, 52). Popper acknowledges that "Kuhn had discovered something [Popper] had failed to see, and [Popper] derived considerable enlightenment from [Kuhn's] discovery . . . [of] what he has called 'normal science' and the 'normal scientist'" (Popper 1974, 1145). Granted, Popper did not approve at all of "normal science," as Kuhn described it. Popper claims that we should pity the normal scientist, and that they have been "badly taught" (Popper 1970/1972, 53). But he recognizes that there are such people working in science.[7]

The novelty of the notion of normal science as distinct from revolutionary changes of theory was a profound contribution to the philosophy of science. Perhaps its greatest impact has been felt in the sociology of science. Much of the work of the Strong Programme in the SSK is concerned with studying the contingencies that influence scientists in their everyday practice (see, for example, Barnes, Bloor and Henry 1996, chapter 2). Even the work of Latour and Woolgar (1979/1986) owes something to Kuhn's discovery of normal science, as does the work by philosophers of science on experimentation, for example, by Ian Hacking and Allan Franklin.

A third important innovation of Kuhn's is the "problem" of scientific revolutions (Kuhn 1997/2000, 312; see also Friedman 2001, 47–49). As we saw above, the notion of scientific revolutions figures importantly in Conant's understanding of science and scientific progress. But it was Kuhn who saw and sought to address the various *epistemic problems* raised by the possibility of scientific revolutions. Revolutionary changes of theory require a radical change in the way scientists see the world and the way they conduct research. Such changes, Kuhn realized, are not apt to be easily accommodated. Rather, they are bound to cause havoc in a research community. And revolutionary changes of theory raise serious challenges for the rationality of science. Kuhn suggested that a revolution involves a change of worldview, and he makes the contentious claim that "after a revolution scientists are responding to a different world" (Kuhn 1962/2012, 111). In his efforts to address the issues raised by revolutionary changes

---

[7] This was a fundamental, unresolved difference between Kuhn's view of science and Popper's view. As Fuller (2004) rightly recognizes, Kuhn thought the success of science was due, to a large extent, to the fact that research is tradition-bound, whereas Popper regarded tradition and dogma as antithetical to the critical spirit that characterizes scientists at their best.

of theory Kuhn took on the thorny issue of progress *through* revolutions, a concern that played no part in Conant's view of science. Indeed, Kuhn's analysis of progress through revolutions gave birth to the notion of Kuhn-loss, a key target of criticism raised by philosophers of science. None of these specific *normative* issues were considered by Conant. His concerns were quite different. Conant was preoccupied with policy issues, which is not surprising given his roles as president of one of the leading research universities in America and as a science advisor to the government during the war years. Kuhn, though, had little interest in policy issues. His concerns were more straightforwardly epistemological. He wanted to answer some of the big pressing questions in the philosophy of science.

Indeed, in the letter Conant wrote to Kuhn shortly after *Structure* was published, Conant singles out the material in the final chapter as especially original and important. Conant notes that he "liked [Kuhn's] last chapter in particular which [he didn't] recall at all from the draft" (Conant 1962). Conant adds that he "couldn't agree with [Kuhn] more heartily when [he says] 'we have to relinquish the notion, explicit or implicit, that changes of paradigm carry scientists . . . closer and closer to the truth'" (Conant 1962).

Fourth, incommensurability is also a concept that has no role in Conant's work. The concept "incommensurability" figures importantly in Kuhn's analysis of the normative dimensions of theory change in *Structure*. Because the concept has generated a vast body of secondary literature, it is worth briefly discussing the role of incommensurability in *Structure* and its fate in Kuhn's later work. "Incommensurability" is like "paradigm." Kuhn used the term in multiple ways, not always noting when and where he was discussing a different issue.[8]

Elsewhere, I have identified four distinct ways in which Kuhn uses the concept "incommensurability" (see Wray 2011, chapter 4). In *Structure* Kuhn uses the term in the following three ways:

i)   to denote meaning incommensurability, the idea that the same theoretical term can have different meanings in different theories, for example Ptolemy's "planets" are not Copernicus' "planets";

ii) to denote topic incommensurability, also called "methodological incommensurability," the idea that competing theories often address

---

[8] Kuhn said many things about incommensurability. But, as Oberheim and Hoyningen-Huene note, not all scholars are agreed on whether there is, in Kuhn's work, a single univocal concept or not. "Some commentators claim that Kuhn's incommensurability thesis underwent a 'major transformation' . . . while others . . . see only a more specific characterization of the original core insight" (Oberheim and Hoyningen-Huene 2018, § 2.3).

different research problems, which explains why scientists committed
to different theories do not always appreciate the strengths of the
competing theory; and

iii) to denote what has come to be called "dissociation," what a historian
of science experiences when they try to understand the science of an
earlier era (see Wray 2011, chapter 4). This was Kuhn's own response
when he read Aristotle's *Physics*.

Later, though, Kuhn would also talk about

iv) the incommensurability between specialty communities: the fact that
physicists, for example, can sometimes have difficulties communicat-
ing with chemists.

It is widely recognized that Feyerabend also began using the term
"incommensurable" at roughly the same time that Kuhn first used it, but
they used the term to quite different ends. Indeed, as Eric Oberheim notes,
their inspiration for the concept was also quite different (Oberheim 2005;
see also Hoyningen-Huene 2005).

Finally, in addition to the concepts of normal science and paradigm, the
problem of scientific revolutions, and the notion of incommensurability,
the fifth important innovation that Kuhn is responsible for is directing
philosophers' attention to the social dimensions of science. Theory change,
on Kuhn's account, is a form of social change (see Wray 2011). Each stage in
Kuhn's cyclical pattern of scientific change, from pre-paradigm science to
normal science, from normal science to crisis, from crisis to revolution and
from revolution to a new normal scientific tradition, is characterized by
a change in the social structure of the research community. The community as a whole is responding to the different challenges that it faces as the
various individual members of the community seek to address the research
problems in their field (see Wray 2011).

Although Conant does discuss the social dimensions of science, especially in his Terry lectures, Kuhn contributes something quite different to
our understanding of the social dimensions of science. When Conant
discusses the social dimensions of science, he is concerned with the importance of "the formation of the scientific societies" in the seventeenth
century, and the consequent "gradual building up of a professional feeling
about science" that altered the way scientists interacted with each other and
discussed their opponents' ideas (see Conant 1947, 7). Conant also recognized the importance of science becoming "a self-propagating social phenomenon," that is, "a tradition" (Conant 1947, 9; 31; 60). These were the

sorts of issues that the sociologist of science Robert Merton discussed. And Conant was very familiar with Merton's research, even using some of it in the General Education Natural Science course (Kuhn 1997/2000, 287).

Some of these considerations figure in Kuhn's social epistemology of science also, but Kuhn actually makes contributions to the *epistemology* of science. For example, he argues that there is a cycle of social changes that scientific specialties move through as those working in a field address research problems. Here we can see Kuhn's commitment to an internalist account of scientific change, a dimension of his project that often goes unnoticed (see Wray 2011, 160–164). Kuhn believed that properly func-tioning scientific research communities respond to changes internal to the community, not to changes in society at large. That is, the dynamical cycle that he presented in *Structure* was driven by conceptual changes, as scien-tists responded to anomalies, not societal factors. This is what makes *Structure* a contribution to the epistemology of science.

In examining the influence of Conant on Kuhn's *Structure* we can better see both where Kuhn's intellectual debts lie, and where his view is most original. Kuhn's debt to Conant, I have shown, is quite extensive, far surpassing any influence that Fleck or Polanyi or Toulmin may have had on his thinking. Importantly, it should be remembered that Kuhn began working with Conant in 1947. And in 1947, Kuhn was only twenty-five years old. Clearly, it should not surprise us that Conant had such a profound influence on him. But I have also shown that there are key elements of Kuhn's view of science that are original, and these are integral to understanding why *Structure* had the impact it had.

A book is more than the sum of the various constitutive claims it asserts. Part of the originality of any book is the way the various constitutive claims are related. I suspect that part of the reason that *Structure* enjoyed the popularity it did, when others were defending similar claims, is due to the way it relates various new ideas that were being discussed widely at the time it was written. The cycle of change that Kuhn presents provides a useful organizing principle for several ideas and themes, a number of which were being discussed by others.

Each of the key contributions that distinguish Kuhn's view from Conant's plays an integral role in the cyclical framework that has made the book so popular. Normal science and paradigms figure importantly in the first part of the cycle, and incommensurability and the problem of revolutionary changes of theory figure in the second part. Finally, Kuhn's focus on the social changes in scientific research communities underlies the whole cycle, shifting our focus away from merely considering the logical relations between evidence and theory, the typical focus of philosophers of science at that time. Adam Timmins rightly notes that the early reviews of Kuhn's *Structure*, especially when compared with the those of Polanyi's *Personal Knowledge*, leave no doubt that Kuhn's book was succinct and clear and provided a provocative image of science (see Timmins 2013, 308–309).[9] No wonder Kuhn's book was such a great success.

---

[9] Indeed, Timmins suggests that one of the chief reasons Polanyi's *Personal Knowledge* was eclipsed by Kuhn's *Structure* was that Kuhn focused on the scientific community whereas Polanyi "tended to focus on the individual scientists" (see Timmins 2013, 315). Apparently Kuhn recognized this difference between his own view and Polanyi's view (see Nye 2011, 244).

CHAPTER 3

Kuhn and the History of Chemistry

The exact nature of the role that the history of science played in Kuhn's thinking when he wrote *Structure* is subject to debate, and Kuhn's views about the relevance of the history of science to the philosophy of science changed over time. Later in his life, he would emphasize the historical or developmental perspective that one learns from history (see Kuhn 1992/ 2000, 112). Further, many have failed to realize which particular episodes in the history of science concerned Kuhn at the time when he was writing *Structure*. Often it is assumed that the Copernican Revolution in astronomy, and the revolutions in twentieth-century physics were the central episodes that he focused on when he was writing *Structure*. Physics, after all, was the field in which he had been trained. And he had already written a book on the Copernican Revolution in astronomy in the mid-1950s.

But a careful examination of *Structure* shows that the history of chemistry played a crucial role as Kuhn worked out his ideas for the book. In this chapter, I examine the influence that the history of chemistry had on the development of Kuhn's philosophy of science. I begin with an analysis of Kuhn's appeals to the history of chemistry in *Structure*. Remarkably, they are quite extensive, and relate to almost all the central themes in the book. Then I turn to an analysis of the sources of Kuhn's reliance on examples from the history of chemistry. The two people who played the most important role in drawing Kuhn's attention to the history of chemistry were (i) his mentor, James B. Conant, and (ii) Leonard Nash, the person with whom Kuhn co-taught the History of Science course after Conant gave it up. Both Conant and Nash were chemists by training. Finally, I examine why philosophers of science have failed to appreciate the extent of the influence that the history of chemistry had on Kuhn's thinking when he wrote *Structure*.

Significantly, Kuhn worked out much of his theory of scientific change through detailed studies of relatively small and esoteric episodes in the history of chemistry, not just through a study of the grand changes in science associated with the likes of Copernicus and Einstein. Granted, these large-scale revolutions, including the so-called "Chemical Revolution" associated with Lavoisier, were important in shaping Kuhn's views. But far too many readers tend to focus exclusively on these "gigantic upheavals," to use an expression Kuhn used in describing the late-eighteenth-century Chemical Revolution.

Chemistry in *The Structure of Scientific Revolutions*

The influence of the history of chemistry on Kuhn's thinking when he was writing *Structure* is quite apparent. Every single chapter of *Structure* includes a discussion or remark about a chemist or a development in chemistry. Of the 127 sources cited in the first edition of *Structure*, approximately 18 percent (23/127) are either (i) sources in the history of chemistry or (ii) scientific reports about findings in chemistry. A perusal of the index in the fourth edition of *Structure* reveals that Lavoisier is the third most discussed scientist in the book. Only Newton and Copernicus are discussed more frequently. And two other chemists, Boyle and Priestley, are among the eight most discussed scientists in the book.[1] Clearly, the history of chemistry figured significantly in Kuhn's thinking about science and scientific change when he wrote *Structure*. Indeed, Paul Hoyningen-Huene has already suggested that "Kuhn's theory is built upon a few historical cases of revolutions, one of which happens to be the chemical revolution," that is, the revolution ushered in by Lavoisier (see Hoyningen-Huene 2008, 114). Hoyningen-Huene's point is that this episode in the history of science "should not be taken as an empirical confirmation of Kuhn's theory," because Kuhn's theory was, in part, derived from this case (see Hoyningen-Huene 2008, 101).

It is worth examining in detail the content of the extended discussions of chemical topics in *Structure*. There are three lengthy discussions that draw on the history of chemistry worth highlighting.

The first occurs in chapter 6, "Anomaly and the Emergence of Scientific Discoveries." For four pages Kuhn discusses the structure of scientific discoveries, focusing on the example of the discovery of oxygen. There is significant overlap here with Kuhn's 1962 paper, "The Historical Structure of Scientific Discovery," published in *Science* (see Kuhn 1962). It appears that his understanding of the complex structure of scientific discovery,

---

[1] Kuhn discusses Boyle not only for his contributions to chemistry, but also for his contributions to pneumatics. But I do not count these latter sources in the figure of 18 percent.

which, he notes, involves both "observation and conceptualization," was significantly shaped by his study of this episode in the history of science (see Kuhn 1962/2012, 55–56). As Kuhn explains, his purpose in the analysis of the discovery of oxygen is to show "how closely factual and theoretical novelty are intertwined in scientific discovery" (see Kuhn 1962/2012, 53).

Kuhn encourages us to give up the misleading locution "oxygen was discovered" (Kuhn 1962/2012, 55). Such an expression "misleads by suggesting that discovering something is a simple act assimilable to our usual . . . concept of seeing" (see Kuhn 1962/2012, 55). Such a view might lead one to think what stands between a scientist and a significant discovery is merely a failure to look in the right direction. Kuhn associates this passive view of discovery with the traditional philosophy of science he seeks to displace. Instead, Kuhn emphasizes the complex structure of discovery, and the fact that scientists must make sense of what they are observing in the process of making a discovery. In his words, "discovering a new sort of phenomenon is necessarily a complex event, one which involves recognizing both *that* something is and *what* it is" (see Kuhn 1962/2012, 55; emphasis in original). Kuhn invokes the anomalous-playing-card experiments of Jerome Bruner and Leo Postman in an effort to make his point vivid to his readers. The subjects of Bruner's and Postman's experiment initially had difficulties seeing the anomalous playing cards as anomalous because of their preconceptions. Instead, they reported seeing normal playing cards – a black four of hearts, for example, would be identified as either a red four of hearts or a black four of spades (see Kuhn 1962/2012, 62–63). Similarly, a number of the chemists who contributed to the discovery of oxygen did not initially see it as oxygen. In his *Science* article Kuhn ties the discovery of oxygen more explicitly to the Chemical Revolution than he does in *Structure*. In that article he claims that, "in the case of oxygen, the readjustments demanded by assimilation were so profound that they played an integral and essential role . . . in the gigantic upheaval of chemical theory and practice which has since been known as the 'chemical revolution'" (see Kuhn 1962, 763–764).

The second sustained discussion of the history of chemistry occurs in chapter 7, "Crisis and the Emergence of Scientific Theories." This discussion runs for three pages. This example from the history of chemistry, that is, "the crisis that preceded the emergence of Lavoisier's oxygen theory of combustion" (see Kuhn 1962/2012, 70), is presented as one of three illustrations of how crisis precedes a change of theory (see Kuhn 1962/2012, 68). Kuhn identifies two factors "of first-rate significance" that generated "a crisis in chemistry" in the 1770s. The first was the "rise of pneumatic chemistry" (Kuhn 1962/2012, 70). Until the mid-1700s, it was still widely believed that "air was the only sort of gas" (Kuhn 1962/2012, 70). But after "Joseph Black showed that fixed air (CO_2) was consistently distinguishable from normal air," others, including Henry Cavendish, Joseph Priestley and Carl Wilhelm Scheele, "developed ... techniques capable of distinguishing one sample of gas from another" (Kuhn 1962/2012, 70). As Kuhn notes, the phlogiston theory was unable to account for the new experimental findings. This led to a proliferation of interpretations of the accepted theory in an attempt to account for the phenomena. Thus, according to Kuhn, "by the time Lavoisier began his experiments on air ... there were almost as many versions of the phlogiston theory as there were pneumatic chemists" (Kuhn 1962/2012, 71). The second source of crisis, Kuhn argues, was the discovery that many bodies when burned or roasted increased in weight, contrary to what the phlogiston theory seemed to entail (see Kuhn 1962/2012, 71). This had led some chemists to postulate or consider the possibility that phlogiston had a negative weight. Again, Kuhn notes that in their efforts to explain the phenomena, chemists were compelled to develop "different versions of the phlogiston theory" (Kuhn 1962/2012, 72). As a consequence, "a paradigm of eighteenth-century chemistry was gradually losing its unique status" (Kuhn 1962/2012, 72). This breakdown of a univocal interpretation of the accepted theory is a symptom of a field in crisis. And as the field of chemistry was in crisis, it was vulnerable to a revolutionary change of theory. The history of chemistry thus provided Kuhn with insight into the role of crisis in the development of scientific knowledge, both signaling and causing a radical change of theory. A crisis is necessary to loosen the grip that the reigning paradigm has on scientists, and make them open to exploring alternative interpretations of the accepted theory, or even a different theory.

The third and final extended discussion of chemistry occurs in chapter 10, "Revolutions as Changes of World View." It runs for five pages. In it Kuhn discusses how a revolutionary change of theory can significantly alter the way scientists understand various laboratory operations (see Kuhn 1962/2012, 129). In this case, it was Dalton's law of fixed proportions that affected the change (Kuhn 1962/2012, 131). Kuhn claims that "examining the work of Dalton and his contemporaries, we shall discover that one and the same operation when it attaches to nature through a different paradigm, can become an index to a quite different aspect of nature's regularities" (Kuhn 1962/2012, 129). That is, scientists will see the world differently depending on which paradigm they bring to their investigation. Kuhn also claims that "occasionally [an] old [laboratory] manipulation in its new role will yield *different concrete results*" (Kuhn 1962/2012, 129; emphasis added). That is, even long-known results are subject to change with a change of paradigm. This analysis of the changes initiated by Dalton is meant to give some concrete sense to the notion that when scientists work with a new paradigm they work in a new world.

According to Kuhn, for chemists who worked in the elective affinity paradigm, which preceded Dalton's paradigm, "the mixture-compound distinction was part ... of the way they viewed their whole field of research" (see Kuhn 1962/2012, 130). Kuhn notes that, although "at the end of the eighteenth century it was widely known that *some* compounds ordinarily contained fixed proportions by weight of their constituents ... no chemist made use of these regularities except in recipes" (see Kuhn 1962/2012, 132). Things were very different once Dalton's paradigm had been accepted. As Kuhn explains, "for Dalton, any reaction in which the ingredients did not enter in fixed proportions was *ipso facto* not a purely chemical process" (see Kuhn 1962/2012, 132–133). That is, the law of fixed proportions was, in some sense, constitutive of chemical processes.

Kuhn argues that as chemists adopted the Daltonian paradigm, chemistry changed in significant, though subtle, ways. For example, "chemists stopped writing that the two oxides of ... carbon contained 56 per cent and 72 per cent oxygen by weight; instead they wrote that one weight of carbon would combine either with 1.3 or 2.6 weights of oxygen" (Kuhn 1962/2012, 133). And, as Kuhn notes, "when the results of old manipulations were recorded in this way, a 2:1 ratio leaped to the eye" (Kuhn 1962/2012, 133). Fixed proportions in chemical processes were now noticed where they had not been noticed before.

Kuhn also notes another significant change. Once chemists approached their research with the Daltonian paradigm, "even the percentage composition of well-known compounds was different. *The data themselves had changed*" (Kuhn 1962/2012, 134; emphasis added). Importantly, Kuhn wants us to see that the shift in paradigm does not just enable scientists to see things they had not noticed before. It even causes scientists to reconsider things they had believed before. And it can even lead to new results. This is an important sense in which scientists working with different paradigms work in different worlds.

Again, it seems clear that Kuhn's main point in this chapter is an outgrowth of his study of this specific case from the history of chemistry. In this respect, his study of that history seems integral to his understanding of science, as it emerged in *Structure*.

In addition to these three extended discussions, Kuhn appeals to the history of chemistry throughout *Structure* as a means to illustrate his many controversial claims about science and scientific change. What follows is just a partial list of examples.

In chapter 1, after Kuhn describes the "defining characteristics of scientific revolutions," he notes that "these characteristics emerge with particular clarity from a study of ... the Newtonian or the chemical revolution" (Kuhn 1962/2012, 6). In chapter 2, Kuhn cites Lavoisier's *Chemistry* as an example of a book that played the equivalent role as a contemporary science textbook, "[expounding] the body of accepted theory, [illustrating] many ... of its successful applications, and [comparing] these applications with exemplary observations and experiments" (Kuhn 1962/2012, 10).

In chapter 4, Kuhn argues that "quasi-metaphysical commitments ... told scientists what many of their research problems should be" (see Kuhn 1962/2012, 41). Here he illustrates his point in a discussion of how the influence of "Descartes' scientific writings" on Boyle and his contemporaries led them to "[give] particular attention to reactions that could be viewed as transmutations" (see Kuhn 1962/2012, 41). This brief discussion in *Structure* draws on Kuhn's 1952 paper on Boyle's chemistry, where the theme is discussed in greater detail (see Kuhn 1952, 22–23).

In chapter 8, Kuhn appeals to the research of Lavoisier to illustrate his point that "often a new paradigm emerges at least in embryo, before a crisis has developed far or been explicitly recognized" (see Kuhn 1962/2012, 86).

In chapter 11, Kuhn appeals to an example from chemistry to illustrate the "reconstruction of history that is regularly completed by postrevolutionary science texts" (see Kuhn 1962/2012, 139). He claims that in many elementary chemistry textbooks, Boyle's definition of an element is presented as part of a narrative of the growth of scientific knowledge as a cumulative process, even though the modern notion of the element is quite different from Boyle's notion (see Kuhn 1962/2012, 140–142). Again, this is an issue that Kuhn had already analyzed in detail in 1952 in his extended study of Boyle's chemistry and its impact on later developments in the field (see Kuhn 1952, 26–27).

In chapter 12, Kuhn appeals to the paradigm introduced by Lavoisier to illustrate what has come to be called "Kuhn-loss." Roughly, Kuhn-loss is a consequence of the fact that "new paradigms seldom or never possess all the capabilities of their predecessors" (see Kuhn 1962/2012, 168; see Post 1971, 229, n. 38, for an early use of the expression "Kuhn-loss"). Kuhn explains that "the transition to Lavoisier's paradigm had ... meant a loss not only of a permissible question but of an achieved solution" (Kuhn 1962/2012, 147–148).

And in chapter 13, Kuhn appeals to the history of chemistry to show how the advocates of the old paradigm deny the progress achieved by the new one. As Kuhn explains, "those who opposed Lavoisier's chemistry held that the rejection of chemical 'principles' in favor of laboratory elements was the rejection of achieved chemical explanation by those who would take refuge in a mere name" (see Kuhn 1962/2012, 162). Discussions of chemical topics are ubiquitous in *Structure*.

Conant, Nash and the Harvard Case Histories

It is worth reflecting on how Kuhn came to draw on so many examples from the history of chemistry in his efforts to illustrate many of the central theses in *Structure*. In the previous chapter, I argued that Conant was a profound source of inspiration for many of the key *concepts* that shaped Kuhn's thinking when he was writing *Structure*. As we saw, Kuhn's focus on conceptual schemes, holism, scientific revolutions and fruitfulness all have their roots in his working with Conant. Here I aim to show that Conant and Nash were principally responsible for Kuhn's engagement with the history of chemistry.

As is widely known, Kuhn's formal training was in physics. In particular, his graduate training was in solid state physics.[2] As we saw earlier, his entry into the history and philosophy of science was through his involvement with the General Education science courses at Harvard, which were an initiative of Conant's (see Kuhn 1984a). As is widely known, even in his capacity as president of Harvard, Conant took an active role in the development and teaching of the General Education science courses (see Hamlin 2016, 284). In fact, Conant "took part for three years in the teaching of [these courses]" (see Bartlett 1983, 98). Conant provides a brief rationale for the General Education history of science courses in both (i) the Foreword to the Harvard Case Histories and (ii) in the Foreword to Kuhn's *The Copernican Revolution* (see Conant 1955, and Conant 1957, xvi–xvii).

---

[2] According to Peter Galison, "Kuhn contributed a significant new approximation method to calculate certain parameters in what was then called solid state physics: the cohesion of energy, the lattice constant, and the compressibility of monovalent metallic solids" (see Galison 2016, 47). Galison adds that Kuhn's thesis work was not "field-changing" (see Galison 2016, 49), and was far removed from contemporary cutting-edge work in quantum field theory (see Galison 2016, 47).

Conant was himself introduced to the study of the history of science in an undergraduate course at Harvard, "Chemistry 8, [a] course largely on 'the historical development of chemical theory'" (Thackray and Merton 1972, 484, n. 29). This was a standard feature of the chemistry curriculum at Harvard. At this time, courses in the history of chemistry were among the most popular courses in the history of science in the United States. In fact, history of chemistry courses accounted for 22 percent of the history of science courses taught at universities and colleges nationwide (see Brasch 1915, 752, table 1, and 757). Thackray and Merton claim that, "at Harvard, as elsewhere, the history of chemistry was godmother to the history of science" (Thackray and Merton 1972, 484, n. 29). Between 1910 and 1915, the course at Harvard was described as having a high level of enrollment, reaching as high as eighty-five students in one year (see Brasch 1915, 752). In fact, Theodore Richards taught the course that year. Richards was the first American to win the Nobel Prize in Chemistry (see Herschbach 2014; also Conant 2017, 52). He was also Conant's father-in-law (that is, after 1920; see Conant 2017, 90).

Not surprisingly, the content of the General Education Natural Science course reflected Conant's own interests and expertise. Consequently, a number of the case studies presented in the course were drawn from the history of chemistry. Ultimately, these became the Harvard Case Histories in Experimental Science. The Harvard Case Histories were the result of Conant's attempt to introduce the case study method of instruction, initially developed by the Harvard School of Law and made popular at Harvard's Business School, into the humanities (see Isaac 2012, 75–80). Central in these cases was the "Chemical Revolution of 1775–1789" that had led to the overthrow of the phlogiston theory (see Conant 1955). In fact, four of the eight Harvard Case Histories in Experimental Science are drawn from the history of chemistry: Vol. 1, *Robert Boyle's Experiments in Pneumatics*; Vol. 2, *The Overthrow of the Phlogiston Theory*; Vol. 4, *The Atomic Molecular Theory*; and Vol. 6, *Pasteur's Study of Fermentation*. Conant edited three of these volumes, and Nash edited the fourth, *The Atomic-Molecular Theory*.

The Harvard Case Histories were not restricted in their influence to courses taught at Harvard. Rather, as Joel Isaac notes, their influence was quite substantial. They "were the final product of a wider project Conant had devised, with the support of the Carnegie Corporation, to provide the materials for college-level courses for lay students of science" (Isaac 2012, 209).

In fact, so great was their impact that some of the particular case studies in the Harvard Case Histories series were enthusiastically taken up and developed further by sociologists and historians of science in the 1970s and 1980s. Boyle's research on pneumatics plays a central role in Steven Shapin's and Simon Schaffer's *Leviathan and the Air-Pump* (see Shapin and Schaffer 1985). And the spontaneous generation debate was discussed in detail by John Farley and Gerald Geison (see Farley and Geison 1974). The attraction of these episodes to sociologists of science was no doubt, due to their focus on scientific practices, something that would not be picked up on by philosophers of science until the groundbreaking work of Allan Franklin and Ian Hacking (see, for example, Franklin 1986 and Hacking 1981).

As noted earlier, Kuhn expressed great admiration for Conant (see Kuhn 1997/2000, 259–260). So, given his involvement with Conant and the General Education Natural Science course, it is not surprising that Kuhn's thinking about scientific change was worked out using examples from the history of chemistry. Even after Conant gave up teaching the course, the course continued to have a strong influence from chemistry as Nash, the person with whom Kuhn subsequently taught it, was a chemist by training (see Kuhn 1997/2000, 281; 283–284; see also Jacobs 2010, 329). As noted already, Nash was responsible for bringing together the volume on the atomic-molecular theory in which Dalton plays a central role (see Nash 1950/1957). And it seems clear that Kuhn relied heavily on Nash's analysis of Dalton's work on subsequent developments in chemistry in his discussion of Dalton in chapter 10 of *Structure*, discussed at length above. So, clearly, as Conant's teaching assistant, and then as heir to the General Education science course, Kuhn was compelled to think about the history of chemistry. No wonder it came to play such a significant role in *Structure*.

There is further evidence that the history of chemistry played a significant role in shaping Kuhn's thinking about science when he was writing *Structure*, and that Nash was personally responsible for the influence. In a letter from Nash to Kuhn, dated April 10, 1957, Nash addresses in detail a variety of questions that Kuhn had about the history of chemistry, questions related to Lavoisier, Dalton, Gay-Lussac and Proust, as well as questions about the secondary literature, specifically, about a recent book by Maurice Daumas and articles by A. N. Meldrum (see Nash 1957). All of these scientists and historians are either discussed or referred to in chapters 6 and 10 of *Structure*. This exchange with Nash was critical for Kuhn as he was working on the book. Kuhn clearly valued both Nash's extensive knowledge of the history of chemistry and his insights. Indeed, in the

Preface to *The Copernican Revolution* Kuhn lists Nash as one of the people who provided feedback on that manuscript (see Kuhn 1957, ix). And in the Preface to *Structure*, Kuhn notes that Nash "was an even more active collaborator [than J. B. Conant] during the years when [Kuhn's] ideas first began to take shape" (see Kuhn 1962/2012, xlv).[3]

In addition, the content of Kuhn's Lowell Lectures further supports the claim that Kuhn was deeply influenced by the history of chemistry during the period when he wrote *Structure*. Indeed, they show how long Kuhn had been thinking about the history of chemistry. These lectures were presented in 1951. Kuhn describes the Lowell Lectures as his first attempt to write *Structure*, and they are filled with examples drawn from the history of chemistry, a number of which would appear in a more developed form in *Structure* (see Kuhn 1997/2000, 289). For example, he discusses John Dalton's chemical atomism in Lecture 3, "The Prevalence of Atoms." Robert Boyle's *The Skeptical Chemist* is discussed extensively in Lecture 4, "The Principle of Plenitude," as is the phlogiston theory. Themes from the history of chemistry are also discussed in Lecture 5, "Evidence and Explanation." Dalton is again discussed in Lecture 6, "Coherence and Scientific Vision," with special attention to what Kuhn there calls "the chemical law of multiple proportions." Kuhn was thus clearly immersed in thinking about the history of chemistry as he developed his general philosophy of science (see Kuhn 1951b). Indeed, Kuhn's first papers in the history of science were on topics in the history of chemistry, specifically on Boyle and Newton (see Kuhn 1997/2000, 290–291). The Boyle paper corrects misconceptions about Boyle's contribution to our understanding of the notion of an element. Further, Kuhn makes clear that during this period he was getting advice and guidance from Nash.

Chemistry, Physics and the *Philosophy* of Science

In the 1940s and 1950s, chemistry was not a common subject of reflection for *philosophers of science*. The Vienna Circle had set the agenda, and their focus was on physics and mathematics (see Edmonds 2020, 23). Insofar as they were trained in the sciences, the Vienna Circle positivists were, for the most part, trained in either physics or mathematics. Hans Hahn's appointment was in mathematics (see Frank 1949/1961, 41–42). Philipp Frank was a physicist (Frank 1949/1961, 42). Moritz Schlick completed a Ph.D. in physics, under Max Planck's supervision (see Edmonds 2020, 16). And

---

[3] Struan Jacobs provides a more comprehensive analysis of the relationship between Nash and Kuhn (see Jacobs 2010, 329–333).

Rudolf Carnap had also studied some physics (on Schlick, see Joergensen 1970, 848). Even Otto Neurath, remembered most for his interests in the social sciences, "began his studies in mathematics and physics in Vienna" (see Edmonds 2020, 12). And Edgar Zilsel, Karl Menger and Kurt Reidemeister had all studied mathematics (see Edmonds 2020, 17).

The Logical Positivists were especially, perhaps narrowly, interested in the developments in modern physics, and principally with the development of the theories of special relativity and general relativity, and quantum mechanics. Even when they escaped from Europe and relocated to America they continued to focus narrowly on physics.[4]

Their focus on physics created an environment that was not particularly amicable to theorizing about chemistry and drawing on examples from its history.[5] Physics, at least those parts of it that concerned the Logical Positivists, was thoroughly mathematized. In contrast, chemistry did not exhibit the formal structure that lent itself to the sort of logical analysis for which the members of the Circle were famous (see Scerri and McIntyre 1997, 216; 221). Indeed, in a comprehensive overview of the relative neglect of philosophy of chemistry by philosophers of science, Eric Scerri and Lee McIntyre note that "it is not clear that the laws of chemistry, if indeed they exist, . . . can be axiomatized" (see Scerri and McIntyre 1997, 216). Scerri and McIntyre also note that "predictions which are made from the so called 'periodic law' do not follow deductively from a theory in the same way in which idealized predictions flow almost inevitably from physical laws, together with the assumption of certain initial conditions" (Scerri and McIntyre 1997, 221). The Logical Positivists were more comfortable with contrived examples involving white swans and black ravens than the sort of real-life, and complicated, examples that one might find in the history of chemistry.

I suspect that at least part of the reason that Kuhn's philosophy of science met the resistance it did was the fact that it was not addressing the sciences that had played a central role in the tradition extending from the Vienna Circle positivists. The First Vienna Circle, Otto Neurath, Hans Hahn and Philipp Frank, read Ernest Mach, Pierre Duhem and Henri Poincaré (see Frank 1949/1961, §§ 3, 4, 7 and 8). These people were also concerned principally with physics and astronomy. Though Duhem wrote

---

[4] George Reisch (2005) notes that the Logical Empiricists who relocated to America were quick to distance themselves from their earlier interests in left-leaning politics. This is especially evident after the war, with the rise of McCarthyism in America.
[5] Indeed, chemistry was not the only natural science neglected by the Logical Positivists. As David Edmonds notes, the Vienna Circle positivists "paid little attention to what became known as genetics and the science of inheritance" (see Edmonds 2020, 23).

extensively about the history of chemistry (see Bensaude-Vincent 2005, 628), his principal work on the philosophy of chemistry, *Le mixte et la combinaison chimique*, published originally in 1902, was not translated into English until 2002 (see Schummer 2006, 26).

Not surprisingly, most philosophers of science in the 1950s and 1960s were not familiar with either chemistry or the history of chemistry. Indeed, this is still true today. And it seems that most did not think that these developments were particularly relevant to developing a general philosophy of science. Joachim Schummer (2006) suggests that "there is a rule of thumb about the philosophers' interest in the sciences: *the smaller the discipline, the more do philosophers write about it*, with the exception of the earth sciences" (Schummer 2006, 21; emphasis in original). Chemists, he notes, produce a body of literature that far exceeds that produced in the other sciences (see 2006, 20, figure 1). But philosophy of chemistry is dwarfed by philosophy of physics and philosophy of biology. The long-standing and persistent commitment to reductionism, which ultimately privileged physics, also likely played a role in the lack of regard given to chemistry (see, for example, Oppenheim and Putnam 1958). This has also been noted by Scerri and McIntyre (1997, 214; see also Schummer 2006, 27).

But Kuhn's research on the history of chemistry was crucial for him, as it was through this study of various episodes in the history of chemistry that he worked out and, ultimately presented, his novel theory of scientific knowledge and scientific change.

Why Have We Not Seen the Influence of the History of Chemistry on Kuhn Before?

One might wonder why historians of the philosophy of science have failed to notice the influence that the history of chemistry had on Kuhn's thinking, especially as he was writing *Structure*. In this section, I want to examine two considerations that offer insight into this issue.

First, one reason that people have failed to notice the influence of the history of chemistry on Kuhn's thinking is because many of the philosophers, historians and sociologists of science who have discussed the development of Kuhn's ideas have tended to focus on explanations in terms of external factors.[6] They have traced the development of his ideas

---

[6] There are exceptions, most notably Paul Hoyningen-Huene's (1989/1993) *Reconstructing Scientific Revolutions*. This book provides a detailed analysis of the development of Kuhn's ideas. But many philosophers writing on Kuhn do not even acknowledge that his ideas changed through the years.

to his cultural milieu, or idiosyncratic experiences in Kuhn's own personal development. Steve Fuller's work is an example of this sort. Fuller argues that Kuhn's view is a consequence of the Cold War culture in which Kuhn began his training as a historian of science (see Fuller 2000). George Reisch also argues that Kuhn was profoundly influenced by Cold War thinking, but Reisch's focus is on the then-popular thinking on the psychology of propaganda (see Reisch 2016; also 2019). John Forrester (2007) has emphasized the influence of psychoanalysis on Kuhn's thinking (see also Andresen 1999). Carlos Pinto de Oliveira has suggested that Kuhn was profoundly influenced by scholarship in the history of art (see Pinto de Oliveira 2017). Some of these studies are quite insightful, but they tend to obscure the fact that Kuhn was reading the history of science extensively, including, and perhaps especially, the history of chemistry. These externalist accounts of Kuhn's development have also obscured the fact that *Structure* is about the natural sciences. Kuhn is quite explicit about this.

Second, I am inclined to think that Kuhn contributed to obfuscating the influence of the history of chemistry on his thinking. He did so in three ways. First, Kuhn's own book-length contribution to the General Education science course at Harvard, his study of the Copernican Revolution in astronomy, obscures the influence of the history of chemistry on his thinking (see Kuhn 1957). Kuhn devoted a lot of time to this study, ultimately including it in his failed tenure application at Harvard (see Fuller 2000, 219–220, n. 90; 383; see also Kuhn 1997/2000, 292).[7] But it is worth remembering that Kuhn's early scholarly publications in the *history of science* were in the history of chemistry. Indeed, Kuhn's first papers in the history of science were studies of early modern chemistry (see Kuhn 1951a and 1952). I have already noted above specific themes and issues in *Structure* that draw on his 1952 paper on Boyle. Kuhn also gives some indication of the importance of these early researches in the history of chemistry to honing his historical sensitivity in the interview he gave in Greece near the end of his life (see Kuhn 1997/2000, 290–291). He notes, for example, that it was in the course of writing his first historical papers on Boyle and Newton that he learned that "sometimes you have to go way back in order to find the starting point, to write something that indicates how powerful ... prior beliefs were and why they ran into trouble" (see Kuhn 1997/2000, 291). This insight was crucial for Kuhn's philosophy of

---

[7] Details of Kuhn's application for a *tenure track appointment* at Harvard are discussed in Hufbauer (2012, 434). And details about his subsequent bid for *tenure* and the review of his application for tenure are discussed in Hufbauer (2012, 439–440).

science. We can only fully appreciate the resistance scientists have to new ideas, new theories, if we first appreciate how they saw the world before the introduction of the new theory.

Second, Kuhn's account of his Aristotle epiphany has probably misled people as well. As we saw in Chapter 1, Kuhn retold the story of the Aristotle epiphany a number of times. For example, in the 1970s Kuhn notes that "I all at once perceived the connected rudiments of an alternative way of reading the texts with which I had been struggling" (Kuhn 1977, xi). Then in the 1980s Kuhn tells us that "[his] jaw dropped, for all at once Aristotle seemed a very good physicist indeed, but of a sort [he had] never dreamed possible" (see Kuhn 1987/2000, 16). And, finally in the 1990s Kuhn reports that he "had wanted to write *The Structure of Scientific Revolutions* ever since the Aristotle experience. That's why [he] had got into history of science – [he] didn't know quite what it was going to look like, but [he] knew the noncumulativeness; and [he] knew something about what [he] took revolutions to be" (Kuhn 1997/2000, 292–293).

No doubt this was a profound experience for Kuhn, and it played a significant role in clarifying his thinking about the nature of scientific change. But the example of the transition from Aristotelian physics to Galilean physics at the core of his epiphany does not play much of a role in *Structure*. Instead, his detailed position was worked out on other examples, including those drawn from the history of chemistry discussed above.

Finally, people have probably overlooked the impact of the history of chemistry on Kuhn's thinking in *Structure* because Kuhn stopped discussing and drawing on examples from that history as he defended and developed his view, in response to the criticism raised against *Structure* after its initial publication. As he developed his stance, he drew on other examples, including: Max Planck's contribution to quantum mechanics, Alessandro Volta's contribution to the study of electricity, the transition from Aristotelian to Newtonian mechanics and the Copernican Revolution in astronomy (see Kuhn 1987/2000). The preoccupation with Planck is largely a consequence of Kuhn's involvement in "the years 1961–1964 with Sources for History of Quantum Physics, an archival project" on the early history of twentieth-century physics. This, in turn, led to his detailed historical study of the black-body problem (see Kuhn 1978/1987, xi). Further, the fact that Kuhn initially trained as a physicist further obscures the influence that the history of chemistry had on his thinking. People readily think about the examples he discusses from the history of physics and astronomy. A number of these examples are quite familiar to philosophers of science because they are the same examples discussed by the Logical Positivists.

Incidentally, historians of the philosophy of science are not the only ones who have failed to see the impact that the history of chemistry had on Kuhn as he was writing *Structure*. It seems some scientists have as well, including chemists. James Marcum notes a number of citations to Kuhn by a number of chemists, including the Harvard chemist George Whitesides (Marcum 2015, 188). Whitesides rightly notes that Kuhn believes that "revolutions occur only when there is no alternative," that is, no alternative to a change in theory (Whitesides 2013, 8). But then Whitesides makes the odd and baseless claim that "Kuhn's argument is based on physics in the period of the early 1900s, when observations in spectroscopy and thermo-dynamics forced the development of quantum mechanics" (see Whitesides 2013, 8). On the basis of this claim, Whitesides then asserts that Kuhn's "formulation of revolutions in science, thus, applies to one part of the history of physics; how broadly it applies to other fields is up for discussion" (see Whitesides 2013, 8). Not only does he not realize the extent to which Kuhn was influenced by the history of chemistry, it appears as if Whitesides never read the parts of *Structure* in which Lavoisier and the Chemical Revolution are discussed.

My aim has been to draw attention to one of the most important, but overlooked, sources of Kuhn's ideas in *The Structure of Scientific Revolutions*. Contrary to the popular trend of focusing on external factors in explaining Kuhn's views, factors related to his social milieu or personal experiences, I have focused on the influence of the books and articles he was reading and thinking about in the history of science; specifically, publications in the history of chemistry. I have argued that there is good reason to think that the history of chemistry had a profound influence on Kuhn's thinking, and what is remarkable is that this influence has eluded our attention for so long. I have also argued that his interest in the history of chemistry was due to the influence of Conant and Nash. Finally, I have noted that the particulars related to the history of chemistry that Kuhn discusses in *Structure* have had little impact on subsequent developments in general philosophy of science.

CHAPTER 4

Kuhn and the Logical Positivists

So far I have emphasized the influence of James B. Conant and the history of chemistry on Thomas Kuhn as he wrote *Structure*. It is worth examining another influence on Kuhn when he was writing *Structure*, specifically the influence of Logical Positivism. *Structure*, after all, was written when Logical Positivism was the dominant philosophy of science in America. In fact, with the mass migration of philosophers from Central Europe in the 1930s, Logical Positivism had more or less displaced pragmatism, the philosophy of science that is often regarded as distinctively American in its origins. Indeed, the influence of the Logical Positivists in philosophy was not limited to the philosophy of science. The Logical Positivists also had a profound influence on the philosophy of language, which had, to a large degree, supplanted metaphysics in the middle of the twentieth century. The Positivists, as well as Ludwig Wittgenstein, had suggested that many metaphysical problems were merely pseudo-problems, a consequence of misuses of language.

Many have remarked on the irony that *Structure* was published as a contribution to the Logical Positivists' *International Encyclopedia for Unified Science*, then under the editorship of Rudolf Carnap and Charles Morris (see, for example, Grene 1963). So, insofar as Kuhn purports to be offering "a decisive transformation in the image of science by which we are now possessed," it is natural to think that the book might contain arguments explicitly directed against Logical Positivism (see Kuhn 1962/2012, 1). But in fact, this is not what we find. There are surprisingly few explicit references to anyone associated with Logical Positivism. Nonetheless, I believe that Logical Positivism did have an impact on Kuhn when he was working on *Structure*. This is most obvious in Kuhn's Lowell Lectures where he explicitly discusses views widely associated with Logical Positivism.

I will begin with a review of what has come to be called the debate about the Kuhn–Carnap connection. This debate concerns the extent to which

Kuhn was influenced by Carnap's later work, which many philosophers regard as quite similar to Kuhn's view. Carnap was, without a doubt, the leading Logical Positivist in the 1950s. I will then analyze what Kuhn actually says about Logical Positivism in *Structure*. Then I will briefly discuss where Kuhn might have got his view of Logical Positivism, that is, the view he aims to challenge or supplant in *Structure*. I argue that there is good reason to believe that W. V. Quine had a significant influence on Kuhn's understanding of Logical Positivism. Finally, I will examine Kuhn's unpublished Lowell Lectures, originally presented in 1951. Kuhn claims that these lectures were his first attempt to write *Structure*. In one lecture, Kuhn critically discusses the type of philosophical analysis of science that the Logical Positivists engaged in. I argue that Kuhn's discussion in the Lowell Lectures suggests that he was more attuned to Logical Positivism than is usually thought. Further, though this discussion was not included in *Structure*, traces of it are in his presentation of the "paradigm" concept, which was only developed some years after the Lowell Lectures, following his stay at the Center for Advanced Study in the Behavioral Sciences at Stanford.

Before proceeding, I note that I will use the terms Logical Positivism and Logical Empiricism interchangeably, following the practice of Alan Richardson and Thomas Uebel (see Richardson and Uebel 2007, 1, fn. 1).

The Kuhn–Carnap Connection

There is a substantial body of literature on the relationship between Kuhn's philosophy and Carnap's, in particular Carnap's later philosophy, which was developed in the 1950s. These reflections are prompted, in part, by a letter Carnap sent to Kuhn after he had read the manuscript of *Structure* which Kuhn had submitted for publication in the *International Encyclopedia of Unified Science*. In the letter, Carnap emphasizes continuities and similarities between his own view and the view Kuhn presents in *Structure* (see Reisch 1991, 266–267). Specifically, Carnap notes that "you emphasize that the development of theories is not directed toward the perfect true theory, but is a process of improvement of an instrument. In my own work on inductive logic in recent years I have come to a similar idea" (see Carnap 1962 in Reisch 1991, 267).

Probably Carnap saw his work in inductive logic as either continuous with or a part of science, for that was an integral part of the Logical Positivists' conception of philosophy (see, for example, Carnap 1932/1959; and Schlick 1930–1931/1959). Carnap continues the above-cited sentence with the following remark: "*my* work and that of a few friends in the step for step solution of problems should not be regarded as leading to 'the ideal system,' but rather as a step for step improvement of an instrument" (see Carnap 1962 in Reisch 1991, 267; emphasis added).

This letter, along with another one from Carnap to Kuhn dated April 1960, was originally published in a paper by George Reisch, with the provocative title: "Did Kuhn Kill Logical Empiricism?" Reisch concludes that Carnap did not perceive Kuhn's book or the view expounded in it as a threat to Logical Positivism. In fact, Reisch argues that

> if [Carnap's] principle of language planning in philosophy is extended to cover choices between radically different theories in science – and ... Carnap makes just such an extension – the picture of revolutionary science that results is very much like Kuhn's. (Reisch 1991, 275)

So, like Carnap, Reisch sees a significant degree of continuity between the Logical Empiricism of the early 1960s, and the view Kuhn presents in *Structure*. Thus, as far as Reisch is concerned, Kuhn did not kill Logical Empiricism.[1] Reisch's paper started a debate that is still ongoing about the Kuhn–Carnap connection. The key question is: How similar is Kuhn's view in *Structure* to the view that Carnap was developing in the 1950s?

Like Reisch, Michael Friedman also notes that "the affinities between Carnap's philosophy of linguistic frameworks and Kuhn's theory of scientific revolutions are ... pervasive" (see Friedman 2003, 20). This has come to be called the revisionist view, as the common narrative in the late 1960s and early 1970s was that Kuhn was part of a movement, the historical school in the philosophy of science, which was a radical departure from Logical Positivism and was responsible for its downfall. The revisionists see far more continuity between Kuhn's philosophy and the so-called Received View, the view of the Logical Positivists.

---

[1] Long before Reisch's paper was published, Popper claimed that it was *he* who had killed Logical Positivism: "Everybody knows nowadays that logical positivism is dead. But nobody seems to suspect that there may be a question to be asked here – the question 'Who is responsible?' or, rather, the question 'Who has done it?' ... I fear I must admit responsibility" (Popper 1974/1992, 99). Popper's false modesty is rather remarkable, given his reputation. And his confession is quite at odds with the popular story. Popper is so often treated as one of the Logical Positivists (see, for example, Longino 1990, 22; Bird 2000, 234). Indeed, recently, David Edmonds emphasizes the fact that, certainly in the early days, Popper did identify with the Logical Positivists. For example, in the 1940s, when he was trying to secure a publisher for *The Open Society and Its Enemies*, Popper "claimed to have been a Circle member" (see Edmonds 2020, 227). And, though recognizing Popper's ambivalence toward the Circle, Hempel's wife suggested that Popper was a member of the Vienna Circle in a letter written to Popper (see Edmonds 2020, 226).

Jonathan Tsou, on the other hand, argues "against the revisionist conclusion that Carnap's and Kuhn's philosophical views are closely aligned" (Tsou 2015, 52). Tsou insists that "Kuhn's philosophical views ... represent a revolutionary departure from Carnap's" (Tsou 2015, 52). Contrary to what the revisionists suggest, Tsou argues that "it is mistaken to regard Carnapian linguistic frameworks as straightforward analogues (or even formal complements) to Kuhnian paradigms" (see Tsou 2015, 57). More importantly, Tsou argues that the "differences in Carnap's and Kuhn's methodological assumptions reflect two radically contrasting styles of doing philosophy of science" (Tsou 2015, 58). Specifically, Carnap was concerned with logical analysis, whereas Kuhn was concerned with historical analysis. This, Tsou claims, is the most fundamental difference. In fact, Tsou argues that "the chief role that *Structure* played in the decline of logical empiricism and emergence of post-positivist philosophy of science was methodological" (Tsou 2015, 61). So, on Tsou's view, it is not Kuhn's substantive view that is his legacy to the philosophy of science, but rather his methodology, his turn to historical analysis as a means to addressing philosophical questions about science.

Others have analyzed Kuhn's relationship to Logical Positivism more generally, that is, as distinct from the relationship between Kuhn's views and those of Carnap. Alexander Bird, for example, argues that, contrary to the widely held view, "Kuhn is not a radical anti-positivist. On the contrary, Kuhn is better seen as firmly belonging to the same tradition as the positivists and empiricists such as Neurath, Carnap and Popper" (Bird 2000, 234). Kuhn's appeal to the history of science has often been seen as an attempt to take philosophy of science in a new direction, that is, to naturalize it. But Bird believes that Kuhn fails to follow through on this promise. Further, Bird attributes the "demise of empiricism" to "the philosophical advances ... that took place after and independently of the publication of *The Structure of Scientific Revolutions*," developments that Bird claims "Kuhn either ignored or explicitly rejected" (see also Bird 2000, 278–279). Specifically, Bird has in mind: "(a) advances in the theory of reference, in particular externalist, non-intensionalist and causal approaches; and (b) advances in epistemology, namely naturalized epistemology and externalism" (Bird 2000, 279).

Gürol Irzik offers another perspective on Kuhn's relationship to Logical Positivism. He suggests that it is a mistake to think that it was Kuhn's principal target. Rather, Irzik argues that "the primary target of SSR is not logical positivism, but the textbook image of science" (see Irzik 2012, 15). It is from textbooks that many scientists and philosophers of science get the idea that the growth of scientific knowledge is strictly cumulative, with no setbacks, a view that Kuhn explicitly argues against.[2]

This debate about Kuhn's relationship to Logical Positivism still continues today.

What Does Kuhn Say in *Structure*?

Kuhn does in fact discuss Logical Positivism explicitly in *Structure*, but only in one chapter, specifically, chapter 12, "The Resolution of Revolutions." Here he presents a view he calls probabilistic verificationism, and which he attributes to Ernest Nagel (see Kuhn 1962/2012, 144–145, see fn. 1). Indeed, Nagel is an apt person to represent the Logical Positivists, at least insofar as any single individual can represent the movement. Nagel, after all, had visited the Vienna Circle in the 1930s (see Nagel 1936a), and was responsible, to some degree, for bringing their views to America (see Ayer 1959, 7; but also Nagel 1936a and 1936b).

Further, Kuhn knew Nagel quite well. In the Preface to *Structure* Kuhn thanks Nagel for giving feedback on a draft of the manuscript (see Kuhn 1962/2012, xlvi).[3] And, in a letter to the editor with whom he was working at the University of Chicago Press, Kuhn cites a supportive passage from his correspondence with Nagel about his manuscript. Nagel wrote that:

> I have read your manuscript with unbounded admiration and personal profit, and have no doubt that its publication will be a major contribution to the literature of the philosophy as well as the history of science. The material you have included for analysis and the issues you have discussed seem to me of greatest importance and I am confident that the book will have a wide audience of readers, not restricted to the professional historians or philosophers. (Kuhn 1961a)

---

[2] Indeed, this might be one of the places where Kuhn was influenced by Ludwik Fleck. Fleck explicitly flags the importance of textbook science for understanding the creation and transmission of scientific knowledge (see Fleck 1935/1979, 112 and 161). Textbook science, Fleck claims, is one of the "four types of science . . . with its own characteristic literature," along with journal science, handbook science and popular science (161).

[3] Kuhn begins by thanking James B. Conant, "who first introduced [him] to the history of science and thus initiated the transformation in [his] conception of the nature of scientific advance." He then thanks Leonard Nash, with whom he taught the history of science course after Conant gave it up, and Stanley Cavell, describing him as "the only person with whom I have ever been able to explore my ideas in incomplete sentences" (Kuhn 1962/2012, xlv). Then Kuhn thanks "the four whose contributions proved most far-reaching and decisive" when he circulated a draft version of the book: Paul Feyerabend, Ernest Nagel, H. Pierre Noyes and John Heilbron (Kuhn 1962/2012, xlvi).

Kuhn's discussion of probabilistic verificationism is part of a larger discussion where he also critically discusses Karl Popper's falsificationism. Kuhn regards his own view of science as an alternative to both (i) probabilistic verificationism, which is a form of confirmation theory, and (ii) falsificationism. These were the two most influential philosophical theories of science at the time (see Kuhn 1962/2012, 8).

Against Popper's falsificationism, Kuhn argues that every theory faces anomalies all the time. Consequently, if scientists were to follow Popper's prescriptions, they would reject all theories. But this is not how scientists behave.[4] Kuhn's treatment of falsification in *Structure* no doubt contributed to the charge that Kuhn failed to keep a clear distinction between normative issues and descriptive issues, a charge he would encounter at the Bedford College conference, and many times afterwards. I will return to this issue later, in Chapter 9.

Against the positivists, Kuhn raises the objection that they misunderstand the nature of verification in science. According to Kuhn, scientists cannot really determine whether a theory is true, as the Logical Positivists seem to suggest. Rather, Kuhn claims, "verification is like natural selection: it picks out the most viable among the actual alternatives in a particular historical situation" (Kuhn 1962/2012, 145).[5] Theory evaluation is comparative and constrained by the theories that have actually been developed at a given time, when the theories are actually being evaluated. Consequently, an inference to the truth of a theory is unwarranted. This is a point that we now associate more often with Bas van Fraassen, and his Argument from a Bad Lot, or Kyle Stanford, and his Argument from

---

[4] In the paper Kuhn presented at the Bedford College conference, he noted his "disagreement with Sir Karl is most nearly explicit" with respect to "[1] my emphasis on the importance of deep commitment to tradition and [2] my discontent with the implications of the term 'falsification'" (Kuhn 1970/1977, 168; numerals added). Specifically, Kuhn claims that the sort of test that concerns Popper "does not occur" in normal science (see Kuhn 1970/1977, 272). Kuhn notes that "both 'falsification' and 'refutation' are antonyms of 'proof.' They are drawn principally from logic and from formal mathematics; the chains of argument to which they apply end with a 'Q.E.D.'; invoking these terms implies the ability to compel assent from any member of the relevant professional community" (Kuhn 1970/1977, 281). Kuhn argues that such a notion of proof is not operative in the natural sciences (see Kuhn 1970/1977, 281).

[5] This is not the only comparison between scientific change and natural selection Kuhn makes in *Structure*. He ends the book comparing progress in science to evolutionary change. "The developmental process described [here] ... has been a process of evolution *from* primitive beginnings – a process whose successive stages are characterized by an increasingly detailed and refined understanding of nature. But nothing that has been ... said makes it a process of evolution *toward* anything" (see Kuhn 1962/2012, 169–170). Like biological evolution, the evolution of scientific knowledge is nonteleological. This aspect of Kuhn's view, though commented on by Carnap, was largely neglected. But it has been discussed more recently by Barbara Renzi (2009), Reydon and Hoyningen-Huene (2010) and Wray (2011), among others.

Unconceived Alternatives (see van Fraassen 1989, 142–143; and Stanford 2006, § 1.3).

Surprisingly, there are no other explicit discussions of Logical Positivism in *Structure*. Indeed, there are very few discussions of the works of philosophers of science in the book. As I have noted elsewhere, in *Structure* "Kuhn cites only 13 philosophical sources (a mere 10% of the total sources cited)" (Wray 2015c, 169). These include just a few key sources in the philosophy of science: Popper's *Logic of Scientific Discovery*, Nagel's *Principles of the Theory of Probability*, Norwood Russell Hanson's *Patterns of Discovery* and Quine's "Two Dogmas of Empiricism." The other philosophical sources Kuhn cites include: Ludwig Wittgenstein's *Philosophical Investigations*, Nelson Goodman's *Structure of Appearance* and Ernst Gombrich's *Art and Illusion*. Though Kuhn's book was intended for philosophers of science, it did not draw extensively on or address, in a detailed manner, the work of other philosophers of science.[6] Its approach, as Tsou noted, was fundamentally different, drawing principally on sources from the history of science.

The Sort of Everyday Image of Logical Positivism

Whether or not Kuhn understood Logical Positivism or whether or not he was up to date with developments in its research program, it is fair to say that he contributed to the popular view (i) that he did not really understand Logical Positivism, and (ii) that the view he targets in *Structure* is a caricature of Logical Positivism.

In an interview in the 1990s, Kuhn addresses the question of where he got "the picture that [he] was rebelling against in *The Structure of Scientific Revolutions*" (Kuhn 1997/2000, 305). Kuhn's answer is a bit surprising. He claims that "I realize in retrospect that I was reasonably irresponsible" (Kuhn 1997/2000, 305). Kuhn describes his target in *Structure* as "that sort of everyday image of logical positivism" (see Kuhn 1997/2000, 306). This remark by Kuhn has been picked up by Alan Richardson and used as a title of an article on Kuhn and the decline of Logical Empiricism (see Richardson 2007, 346).

In fact, Kuhn also suggests that the image of science that became his foil in *Structure* was probably reinforced by what he was reading in the

---

[6] Though Kuhn intended *Structure* to be a contribution to philosophy of science, after the book was published, he was somewhat perplexed about how to characterize it. This is evident from Kuhn's Lecture Notes from 1966, which I discuss in Chapter 11, below (see Kuhn 1966b).

philosophy of science in the mid-1940s. He mentions that while working at the Radio Research Laboratory he had time to read. And he specifically mentions reading "Bertrand Russell's *Knowledge of the External World* ... some von Mises ... Bridgman's *Logic of Modern Physics* ... some Philipp Frank ... [and] a bit of Carnap, but not the Carnap that people later point to as the stuff that has real parallels with me" (Kuhn 1997/2000, 305–306). Indeed, it may have been as late as 1983 before Kuhn became aware of Carnap's work from the 1950s, that is, the work at the center of the debate about the Kuhn–Carnap connection. In a letter from 1983 to Ernan McMullin, Kuhn expresses gratitude to McMullin for drawing his attention to "Carnap's 1956 article" (see Kuhn 1983b). I suspect Kuhn has in mind Carnap's "Empiricism, Semantics and Ontology," originally published in 1950 but reprinted in 1956, for McMullin cites this paper in his 1982 PSA paper, "Values in Science" (see McMullin 1982). It is in that paper that Carnap draws the distinction between internal and external questions of existence. The question "Do atoms exist?" is an internal question, and it is to be decided on the basis of an empirical investigation. The question "Should scientists work with a framework that invokes atoms?" is an external question, and, according to Carnap, it is a question about whether such a framework is useful, given the aims of scientists. This sort of distinction would figure more significantly in Kuhn's later work.

In fact, I think Kuhn is being a bit disingenuous when he suggests that he knew so little about Logical Positivism. Later in that same interview, he notes that

> the philosophy I knew and had been exposed to, and the people in my environment to talk to, were all of them out of the English logical empiricist tradition, in one way or another. This is the tradition which by and large had no use for the continental and particularly the German philosophical tradition. (Kuhn 1997/2000, 321)

But, still, a case can be made that Kuhn was in fact not up to date on the recent developments in the research program of the Logical Positivists. He does, after all, claim not to have read Carnap's more recent work at the time when he wrote *Structure* (see Kuhn 2000, 306). And his own understanding of Logical Positivism may in fact have been shaped significantly by his interactions with Quine. In fact, Kuhn mentions that while he was at the Harvard Society of Fellows, "Quine opened for [him] the philosophical puzzles of the analytic-synthetic distinction" (Kuhn 1962/2012, xli). Kuhn cites Quine's "Two Dogmas of Empiricism," which is without a doubt an attack on the Logical Positivists, and specifically philosophical views that Quine attributes to Carnap (see Quine 1951). Quine, though, has been accused of misreading Carnap. As Richard Creath explains,

> it is to Quine that we owe the widespread conviction that Carnap is some sort of British empiricist whose driving concern in the *Aufbau* ... was the ontological reduction of science to an absolutely certain domain of sense data ... It is also to Quine that we owe the general impression that Carnap's discussion of analyticity is likewise motivated by the quest for certainty and that an analytic sentence is one that will be held true come what may. This is squarely at odds with Carnap's conventionalism ... Still, Quine's caricature is endlessly repeated by others. (Creath 2007, 334–335)

It is quite plausible that Kuhn's conception of Logical Positivism was shaped by what Creath calls "Quine's caricature" of Carnap. Kuhn's interactions with Quine were not limited to their time at Harvard and the Society of Fellows. Quine was also a visitor at the Center for Advanced Study in the Behavioral Sciences in 1958–1959, when Kuhn was there, trying to complete a draft of *Structure*.[7]

In fact, this is not all that Kuhn learned from Quine. During the same interview, Kuhn remarks that Quine's *Word and Object* was also important to him, especially as he wrestled "with the problem of meaning" (Kuhn 1997/2000, 279). Kuhn specifically claims that *Word and Object* and "Two Dogmas of Empiricism" led him to realize that he "didn't have to ... [look] for necessary and sufficient conditions" (Kuhn 1997/2000, 279). The sort of family-resemblance type of thinking that figures in the application of paradigms to new problems thus also probably traces its origins to Quine, and not Wittgenstein, contrary to what many may think. Scientists, Kuhn claims, do not need to reason by logical rules in their efforts to solve outstanding research problems. Instead, they see patterns and similarities between problems (see Kuhn 1970/2000, 169–171; also Hacking 2016).

In fact, the interactions between Kuhn and Quine seem to have been mutually beneficial. In the Preface to *Word and Object*, Quine lists Kuhn, along with nineteen others who had provided "advice and criticism" as he worked on the book. So Kuhn had read a draft of *Word and Object*, and of course "Two Dogmas," long before Stanley Cavell drew his attention to Wittgenstein.[8]

---

[7] Quine was then working on *Word and Object* (Quine 1960, x). Incidentally, Clifford Geertz was also at the Center that year. Kuhn would later interact with Geertz when they were both at the Institute for Advanced Studies at Princeton (see Kuhn 1997/2000, 320).

[8] Some think that Kuhn's appeal to the notion of family resemblance is rooted in Wittgenstein. Though that specific phrase may have originated in Wittgenstein, Kuhn was already thinking along

Indeed, Kuhn continued to engage with Quine's philosophy through-out his career, right up until the end. In the Preface to *Essential Tension*, published in 1977, for example, Kuhn notes that Quine's work had led him to rethink the notion of "meaning change", as well as the incommensur-ability that Kuhn believed divided proponents of competing theories (see Kuhn 1977, xxii–xxiii). There are even letters from 1993 between Kuhn and Quine in the archives at MIT, in which Kuhn attempts to clarify Quine's view, which he intended to address in his uncompleted book, *The Plurality of Worlds*.

Alternatively, some might think that Kuhn could have been influenced in his reading of the Logical Positivists by Popper. In fact, Kuhn had met Popper in 1950 when Popper gave the William James Lectures at Harvard (see Kuhn 1997/2000, 286). But it is quite doubtful that Popper is respon-sible for Kuhn's interpretation of Logical Positivism for two reasons. First, Kuhn's criticisms of Logical Positivism in *Structure* are somewhat tangen-tial to Popper's concerns. Kuhn had said the Positivists failed to see that theory choice is like natural selection, insofar as scientists are choosing the best of the available alternatives, just as natural selection operates on the varieties that have in fact been developed. Thus, it seems that Kuhn did not see the Positivists through Popper's eyes. Second, Popper's William James Lectures were on the topic of "the methodology of social science," a topic tangential to the research interests of most of the Logical Positivists (see Popper 1974/1992, 146–147).[9] What stuck with Kuhn from this time was the fact that "Popper was constantly talking about how the later theories embrace the earlier theories," a view that Kuhn thought was mistaken (see

those lines before he encountered Wittgenstein's writings. Paul Hoyningen-Huene makes a compelling case that it was Cavell who drew Kuhn's attention to the relevance of Wittgenstein's writings sometime *after* April 1961, when Kuhn completed the manuscript that Hoyningen-Huene has christened *Proto-Structure*, and the writing of the final manuscript of *Structure* that was sent to the publisher (see Hoyningen-Huene 2015, 188). Kuhn's appeal to the notion of family resemblance was, as one would expect, in the context of arguing against the need for knowledge of the necessary and sufficient conditions for concept application (see Kuhn 1962/2012, 45). Kuhn was attuned to this issue from his reading of Quine. Hoyningen-Huene notes that Cavell is also responsible for Kuhn's cryptic remarks about the context of discovery and the context of justification in chapter 1 of *Structure* (see Kuhn 1962/2012, 8–9; and Hoyningen-Huene 2015, § 13.2.2).

[9] Kuhn mentions that he was "introduced to Popper at a fairly early stage [in his visit] and we saw a little bit of each other" (Kuhn 1997/2000, 286). Recalling this time in his autobiography, Popper lists the various people he met during what was his first visit to America, and he does not mention Kuhn at all. Specifically, Popper mentions: Quine, C. I. Lewis, Donald Williams and Morton White, from the Philosophy Department; "a number of old friends: the mathematician Paul Boschan, Herbert Feigl, Phillip Frank (who introduced me to the great physicist Percy Bridgman, with whom I quickly became friends), Julius Kraft, Richard von Mises, Franz Urbach, Abraham Wald, and Viktor Weisskopf"; and Gottfried van Haberler, George Sarton, I. B. Cohen and James B. Conant (see Popper 1974/1992, 147).

Kuhn 1997/2000, 286). Kuhn, after all, was insistent that because of the role of radical theory change in scientific development, the growth of scientific knowledge was not strictly cumulative, as Popper's view implies. But Kuhn did acknowledge that, during this visit, Popper recommended that he read Emile Meyerson's *Identity and Reality*, which proved to be very useful to Kuhn (see Kuhn 1997/2000, 286–287).

The Lowell Lectures

Kuhn claims that the Lowell Lectures were his first attempt to write the book he had wanted to write, the book that would become *Structure* (see Kuhn 1997/2000, 289). These lectures were given in 1951. There is significant overlap between some of the themes and issues discussed in them and the themes and issues discussed in *Structure*. For example, it is striking how much of the material on the history of chemistry that Kuhn works into *Structure* was already in the Lowell Lectures. As argued in the previous chapter, the history of chemistry was very important to Kuhn as he worked out his ideas.

But more important for our purposes here is the fact that he devoted a lecture to discussing logical studies of science, the sorts of studies associated with the Logical Positivists. More precisely, Kuhn spends a lecture discussing the value of developing formal languages in science. This seemed to be an integral part of the Logical Positivists' program. Kuhn had an interesting critical insight about the Positivists' views on formal languages and science, as he understood them. This insight was not included in the final version of *Structure*, at least not in a form similar to that which he presents in the Lowell Lectures. But Kuhn did not forget or wholly set aside this insight. Remnants of his reflections on the limitations of formal languages would be merged with his nearly all-embracing concept of "paradigm." It is worth examining what Kuhn said on this issue in the Lowell Lectures.

In Lecture VII, "The Role of Formalism," Kuhn discusses what role formal languages play in science. It is quite remarkable that Kuhn would attempt such a lecture in a nonacademic setting. Granted, those in the audience at the Boston Public Library were likely, indeed, almost certainly, university educated. But this is hardly a topic that would make for an engaging public lecture. In this lecture on formal languages, Kuhn discusses, for example, alternative geometries, that is non-Euclidean geometries, as well as logical connectives, for example, "and" represented as "•" (see Kuhn 1951b, VII – 3 – 2; and

VII – 3 – 5).[10] Though there are not explicit references to the works of the Logical Positivists, given the topics discussed, this lecture seems explicitly aimed at evaluating their program. In the lecture, Kuhn claims that "the ultimate outcome of this program would be to separate entirely the empirical portion of our knowledge from the portion due to the conventions we employ in . . . formalizing it" (see Kuhn 1951b, VII – 6 – 4).

As the lectures are still unpublished, it is worth presenting some of the content in detail. In the lecture, Kuhn makes the following remarks:

- "The attempt to formalize limited areas of existing scientific knowledge has been and will continue to be a useful tool of research" (Kuhn 1951b, VII – 6 – 4).
- For example, "physics profited by the formalization of geometry" (Kuhn 1951b, VII – 6 – 4).
- Specifically, the formalization of geometry in the 1800s "resulted in the recognition of the existence of many geometries different from Euclid's, and this recognition increased the freedom of scientific thought about the characteristics of physical space" (Kuhn 1951b, VII – 7 – 1).
- "Here clearly is an example of a scientific situation in which the separation of empirical and formal elements of knowledge provided the scientist with increased freedom of conceptualization" (Kuhn 1951b, VII – 7 – 1).
- "The Euclidean formalism which had previously been thought necessary was shown to be only one of a number of alternate formalisms, and the scientist was directed to new aspects of experience in an effort to choose among possible alternates" (Kuhn 1951b, VII – 7 – 1).
- In this way, "the application of formalism to a limited area of knowledge as a tool which by creating greater conceptual freedom aids in the resolution of crisis states in individual sciences" (Kuhn 1951b, VII – 7 – 2).

Thus, as these passages from the Lowell Lectures indicate, Kuhn had some appreciation for the Positivists' program, and the value it might have for science.

Kuhn, though, describes himself as "considerably less optimistic about the possibility of completing this research effort than the majority of

---

[10] The unusual notation I use for the citations to the Lowell Lectures is Kuhn's own, which he employs throughout the manuscript of the lectures. The lectures have not yet been published but are in the T. S. Kuhn Archives at MIT. I will use Kuhn's notation for the benefit of other scholars aiming to study the lectures.

contemporary students" (see Kuhn 1951b, VII – 6 – 4). He raises two concerns, one practical and one normative. Let us consider each in turn.

First, he is skeptical that scientists could ever develop a formal language comprehensive enough for all of science. As he explains, the project of developing such a language "calls for the construction of a language adequate to deal unambiguously with infinite variety presented by the totality of our perceptual experience" (see Kuhn 1951b, VII – 7 – 7). And this, Kuhn claims, "is clearly impossible" (see Kuhn 1951b, VII – 7 -7).

But setting this practical problem aside, Kuhn has a concern about the *desirability* of developing such a language. That is, he thinks that pursuing the project would be detrimental to science. Specifically, he argues that such a language would actually be an impediment to the advancement of science.[11] He argues that formal languages leave no room for new concepts. That is, formal languages are rigid and lack the sort of flexibility that scientists need in their efforts to solve outstanding or unresolved research problems. As Kuhn explains:

- "if one were to commence such a project of complete formalization, one would undoubtedly start with the language employed for existing scientific knowledge" (Kuhn 1951b, VII – 8 – 1).

But this will quickly prove to be impractical and problematic, Kuhn explains.

- "Supposing the project completed, its results would be to freeze scientific attention upon just those aspects of nature which are embraced by contemporary science" (Kuhn 1951b, VII – 8 – 1).

Thus, the resulting formal language would be limiting in a stifling way, right from the start. It would not be able to effectively serve scientists in the pursuit of their research goals. Because the language would be developed to capture our contemporary scientific understanding, there would be a built-in conservativism; one that would prove problematic for accommodating new discoveries, specifically, radical new discoveries that depend on new concepts.

- Though "it would provide a place in its meaning system for aspects of nature now considered technically relevant," it would provide "no place for others" (Kuhn 1951b, VII – 8 – 1).

---

[11] Eric Oberheim rightly notes that Feyerabend had a similar preoccupation with "conceptual conservativism." Oberheim characterizes "Feyerabend's early philosophy ... [as] a series of critiques of various forms of conceptual conservativism" (Oberheim 2005, 386).

- "As a result it would not be a language adequate to embrace *new conceptual developments in science*" (Kuhn 1951b, VII – 8 – 1; emphasis added).

This, I believe, is an important insight. Kuhn was especially concerned with what has come to be called science-in-the-making, not just settled science. Indeed, this is one important respect in which Kuhn's concerns were more similar to those of sociologists than to those of philosophers of science. Sociologists as different as Bruno Latour and Stephen Cole were focused on understanding science at the research frontier, where there may not be a settled consensus (see Latour 1987; and Cole 1992).

There is no argument like this presented in *Structure*. That is, nowhere in *Structure* does Kuhn take on the issue of the limitations of formal languages, not even in his discussion of probabilistic verificationism, the view he attributes to Nagel, as a representative Logical Positivist. But significantly, Kuhn does address the issue of the development of new concepts and conceptual understandings in *Structure*. This is where the notion of "paradigm" comes in. The sort of creative flexibility that he suggests would be compromised by a fully developed formal language of science is discussed in *Structure* in relation to the interpretation of paradigms or exemplars.

One of the things that makes paradigms so useful to scientists is that they are malleable, and thus afford insight into scientific problems that have hitherto defied a solution. This is why he insists on the priority of paradigms in chapter 5 of *Structure* (see Kuhn 1962/2012, chapter 5). As Kuhn explains at the end of chapter 4, "Normal Science as Puzzle-Solving," though "normal science is a highly determined activity . . . it need not be entirely determined by rules" (see Kuhn 1962/2012, 42). Logical analyses of science, it seems, treat scientific reasoning as *determined by rules*. The sort of flexibility that Kuhn claimed was threatened by the employment of formal languages was something that Kuhn saw as integral to science and scientific change. It provides scientists with the resources to creatively solve research problems. The various parameters of an exemplar can be adjusted to solve the problem at hand.

What changes between the Lowell Lectures and the *Structure of Scientific Revolutions* is that he no longer explicitly contrasts the needed flexibility with the rigidity of formal languages of the sort associated with the Positivists. But Kuhn certainly regarded the paradigm concept as a radical departure from the Positivists' logical analyses of science.

In fact, Kuhn returns to these issues, discussed in the Lowell Lectures, after the publication of *Structure*. For example, at the

Bedford College conference, he noted that "though logic is a powerful and ultimately an essential tool of scientific enquiry, one can have sound knowledge in forms to which logic can scarcely be applied" (Kuhn 1970/1977, 285). Thus, as far as Kuhn is concerned, a logical analysis of science, the sort of analysis that the Logical Positivists aimed to develop, would only be so illuminating. Aspects of scientific knowledge, specifically, those embodied in paradigms or exemplars, would probably go unnoticed.

In summary, though the Logical Positivists make only brief appearances in *Structure*, their impact on Kuhn's thinking when he was writing the book was quite significant. His mode of presenting his new view of science changed significantly between the early 1950s, when he prepared the Lowell Lectures, and 1962, when he completed the book. He had acquired a new set of concepts, including paradigm and normal science, in the intervening years. And these provided a framework for Kuhn's presentation of his accounts of scientific change and scientific knowledge. And, insofar as the Logical Positivists are addressed in *Structure*, the views he attributes to them may owe a lot to Quine, who had an enormous impact on the reception of Logical Positivism in America. But it is likely that Kuhn had at least two different foils in mind as he wrote *Structure*: (i) the textbook image of science; and (ii) the popular Quinean reading of Logical Positivism.

The Unexpected Uptake: Kuhn and the Social Sciences

In Part II, I examine the uptake to Kuhn's *Structure* in the social sciences. *Structure* was intended to describe the pattern of change and development in the *natural sciences*. Kuhn was quite explicit about this. But much to his surprise, social scientists were profoundly influenced by his analyses of scientific change and scientific knowledge. A number of the concepts presented in *Structure*, most notably, paradigm and paradigm change, were adopted and applied widely in the social sciences.

The most significant impact of Kuhn's work in the social sciences, though, was on the sociology of science. Kuhn set the sociology of science in a new direction. Inspired by Kuhn's account of science, sociologists of science began to examine how social factors affect the content of science. Kuhn was quite surprised, indeed, at times, dismayed, by the developments in the sociology of science that were alleged to follow from his views.

CHAPTER 5

Kuhn's Influence on the Social Sciences

Perhaps no twentieth-century philosopher has had a greater influence on the social sciences than Thomas Kuhn. And perhaps no group read *The Structure of Scientific Revolutions* with more enthusiasm than social scientists (see Kuhn 1997/2000, 307; also Bornmann et al. 2020, 1066, table 7). This is ironic, because in *Structure* Kuhn set out to develop an understanding of the nature of scientific knowledge and scientific inquiry in the *natural* sciences. He did not discuss the social sciences at all, except to remark in the Preface that it was while working at the Center for Advanced Study in the Behavioral Sciences, surrounded by social scientists, that he had hit upon the "paradigm concept," and to note that paradigms are largely absent in the social sciences (see Kuhn 1962/2012, xlii). But with its many suggestive metaphors, *Structure* resonated with social scientists.

In this chapter, I examine Kuhn's influence on the social sciences. His influence on the social sciences was quite pronounced. Much of the language Kuhn uses to characterize the epistemic culture of the natural sciences has been adopted and applied widely by social scientists to their own disciplines. In fact, Kuhn once remarked that "if you go through college in science and mathematics you may very well get your bachelor's degree without having been exposed to the *Structure of Scientific Revolutions*. If you go through college in *any* other field you will read it at least once" (Kuhn 1997/2000, 282–283). *Structure* has even been included on Ph.D. reading lists in some social sciences (see Walker 2010, 434).

On the one hand, the book provided social scientists with a new perspective on their own fields of study as well as new questions about their fields. Many social scientists were concerned with understanding the significance of the alleged differences between their own disciplines and the natural sciences. Others were intrigued by developing a better understanding of the dynamics of conceptual change in the social sciences. Kuhn's book provided a fresh framework for addressing these questions. On the other hand, the influence that *Structure* has had on the social sciences has sometimes been rather unfruitful, some might even say pernicious. Many of the adaptations of Kuhn's views by social scientists are based on misunderstandings of his views. Still, it is undeniable that Kuhn had a significant impact in the social sciences. It is worth examining the influence Kuhn's *Structure* had on the social sciences in detail.

The Impact of *Structure* on Psychology

Before I begin my main discussion of the impact of *Structure* on the social sciences, I want to make some brief remarks about Kuhn's impact on psychology. Psychologists responded to *Structure* in a way quite different from that in which sociologists, political scientists, economists and anthropologists responded to it. This is not surprising. The relationship between psychology and these other fields is somewhat unclear. Sometimes psychology is assumed to be a social science, but at many universities, the department of psychology is not even housed with the departments of sociology, political science, economics and anthropology. These latter fields are clearly social sciences, and I will discuss them together as a group following this brief discussion of psychology. Psychology, on the other hand, is sometimes regarded as a natural science.

David Kaiser has already provided a detailed analysis of the reaction of psychologists to *Structure*. Psychologists, as he notes, were quite enthusiastic about the book. In fact, there are more letters from psychologists in the Kuhn Archives at MIT inquiring about *Structure* than there are from scholars in any other discipline (see Kaiser 2016, 72; 83; and 84, table 4.1).[1] But it is interesting to note that psychologists were most often interested in the psychological implications of Kuhn's remarks on perception and cognition (see Kaiser 2016, 85). That is, they engaged with the remarks that Kuhn made about the theory-ladenness of observation. Kuhn had both illustrated and supported the claims he made about the theory-ladenness of perception by citing or discussing the work of psychologists.

---

[1] It seems that psychologists' interest in *Structure* was more immediate, but then dropped off more quickly than the interest shown by sociologists, political scientists and economists. As Kaiser reports, whereas psychologists account for 18.7 percent of Kuhn's correspondents between 1962 and 1967, they account for only 14.7 percent of his correspondents between 1962 and 1981. In contrast, whereas sociologists, economists and political scientists account for 11 percent of Kuhn's correspondents between 1962 and 1967, they account for 16.4 percent of his correspondents in the longer period, 1962–1981 (see Kaiser 2016, 84, table 4.1).

He cites Jean Piaget and various Gestalt psychologists. He also discusses Bruner and Postman's anomalous-playing-card experiment in some detail.

Psychologists responded to Kuhn's book as if it were a book *about* psychology. In fact, after reading the book, in 1962, the year the book was published, the distinguished psychologist Edwin Boring wrote to Kuhn that "it seems to me probable that the publisher [of *Structure*] does not know that this is a book about psychology" (Boring 1962, cited in Kaiser 2016, 83).

But as Kaiser notes, "Kuhn was not formally trained in psychology, nor were his academic interests limited to that field. Far from it – he read eclectically from his early years" (Kaiser 2016, 89). Indeed, Kuhn's appeal to *the research literature in psychology* is rather selective, and not systematic. That is, Kuhn's citations of research in psychology do not constitute a literature review in accordance with the disciplinary norms of psychology. Rather, they are perhaps best described as eclectic. Consequently, it is somewhat surprising that psychologists regarded the book as a contribution to their field.

Interestingly, unlike sociologists, political scientists, economists and anthropologists, psychologists were generally not led to reflect on the dynamics of change in and the status of their own discipline. In contrast, as we will see shortly, this is precisely how sociologists, political scientists, economists and anthropologists responded to *Structure*. They regarded it as metascientific, offering insight into their own discipline from a perspective outside of their discipline.

Given the differences between psychology and these other fields, for the remainder of this chapter, whenever I refer to the social sciences, it should be understood that I am not including the field of psychology except in those few cases where I explicitly say so.

Kuhn's *Structure*

Kuhn's influence on the social sciences is due, almost exclusively, to the publication of *The Structure of Scientific Revolutions* (see Kuhn 1962/2012). Both Kuhn's earlier historical work, including *The Copernican Revolution*, and his later work, including the essays in *The Road since Structure*, have had a negligible impact on the social sciences. Because of the influence of *Structure* on the social sciences, it is worth briefly outlining the main themes in the book, highlighting those that caught the interest of social scientists.

Structure presents a theory of scientific knowledge and change. As we saw earlier, the view it presents is meant to be contrasted with a popular positivist view of science. The positivists were alleged to maintain that the growth of scientific knowledge is cumulative, without interruptions or setbacks of any significance. Scientists are guided by a method that, in conjunction with careful attention to data, ensures that when they are confronted by a choice between competing theories, they have the ability to unequivocally determine which theory is the superior theory, that is, which theory is closest to the truth. Kuhn's account of science is a reaction to this particular view. Whether or not anyone actually held the popular positivist view is beside the point. It is the view that Kuhn addresses.

Kuhn notes that during his training as a physicist he uncritically accepted this positivist view of science. But then he began to look at the history of science and he started to see things quite differently. Contrary to what the popular positivist view suggests, the growth of science is not a continuous march closer and closer to the truth. Instead, periods of quite rapid growth are interrupted by revolutionary changes of theory. These revolutionary changes of theory are disruptive, and often lead to a loss of putative knowledge in some areas. Further, Kuhn did not believe that there was anything like a scientific method that could unequivocally resolve disputes between proponents of competing theories. Instead, looking at the historical record, Kuhn came to believe that scientists were rather dogmatic, uncritically accepting the theory they were taught. Kuhn did not, however, intend this observation as a criticism of science. Rather, he believed that the dogmatism that characterizes science explains why scientists are so successful in accomplishing their research goals. When scientists uncritically accept the theory they were taught, it gives them the determination to make nature fit the conceptual boxes or categories that the accepted theory provides. The constraints supplied by the accepted theory provide scientists with the focus to resolve challenging research problems, to solve the puzzles of normal science. Only in times of crises, when anomalies persistently resist normalization, do scientists begin to consider alternative theories. And such innovations are often initially met with considerable resistance.

Kuhn also claims that scientists working in the same scientific field do not necessarily share a common understanding of the accepted theory, nor are they bound together by an explicit set of methodological rules. Rather, it is the paradigms or exemplars accepted in a field that bind the scientists working in that field together. Exemplars are widely accepted solutions to research problems, solutions that provide insight into solving related research problems. These exemplars play a crucial role in the socialization of young scientists-in-training. Laboratory exercises are designed to aid students in learning to apply exemplary solutions to other related scientific problems. In fact, Kuhn believes that scientists learn the key concepts of their field by solving problems in the laboratory, not through a close study of a theory. For example, it is in modeling a solution to a problem in a physics lab that a student learns the concepts of mass, velocity, momentum and the like.

Kuhn describes competing theories as providing scientists with incommensurable worldviews. Competing theories do not address precisely the same research problems and they often use key theoretical terms in different ways, with different connotations and denotations. Because of the incommensurability of competing theories, a change of theory in a field can be quite complicated. Scientists working with competing theories may not fully understand each other. Nor are they apt to agree about which scientific problems in their field are most significant. Thus, incommensurability can pose a significant challenge to theory change. At times, it can even *seem* to threaten the rationality of theory change.

Ultimately, when a new theory is deemed to be superior to a long-accepted one, a revolutionary change of theory occurs, and the older theory is replaced by the new theory. Then, a new period of normal science begins, one that is structured around the newly accepted theory, and one in which scientists work with different exemplars or paradigms. Kuhn even suggests that when there is a change of theory, it is as if the scientists in a field are working in a new world afterward. Hence, the scientists who live through such changes see the world differently after the change of theory.

According to Kuhn, this developmental cycle – from normal science to crisis, which is ultimately resolved by a revolutionary change of theory, which initiates a new phase of normal science – characterizes the growth of scientific knowledge in the natural sciences. And as far as Kuhn was concerned, there is no end to the cycle. Every theory will ultimately be replaced by some alternative theory.

Initial Reactions to *Structure*: Some Archival Evidence

Structure attracted the attention of social scientists right from the beginning. This was somewhat unexpected, as Kuhn was claiming to provide an account of scientific change for the *natural sciences* only (see Kuhn 1997/ 2000, 307). Most of the historical examples in *Structure* are drawn from chemistry, physics and astronomy, and none of the historical examples come from the social sciences. Still, *Structure* resonated with social scientists from across the disciplines.

As noted above, Kuhn had little to say about the social sciences in general, other than claiming that he had discovered the "paradigm" concept when he was working along-side social scientists at the Center for Advanced Study in the Behavioral Sciences. But he makes one further interesting remark, at the beginning of the chapter in *Structure* on progress in science. Without going into any details, Kuhn suggests that economics is on a developmental path similar to the developmental cycle he described in *Structure*. He notes that "to a very great extent the term 'science' is reserved for fields that do progress in obvious ways" (Kuhn 1962/2012, 159). Kuhn then suggests that "nowhere does this show more clearly than in the recurrent debates about whether one or another of the contemporary social sciences is really a science" (1962/2012, 159). He compares these debates to the debates that he says characterized the natural sciences in their pre-paradigm stage. Kuhn suggests that these debates about progress and the scientific status of the social sciences "will cease to be a source of concern ... when the groups that now doubt their own status achieve consensus about their past and present accomplishments," as happened in the natural sciences (see Kuhn 1962/2012, 160). He then remarks that "it may ... be significant that *economists* argue less about whether their field is a science than do practitioners of some other fields of social science" (1962/2012, 160; emphasis added).[2]

The Thomas S. Kuhn Archives at the Massachusetts Institute of Technology (MIT) contains a vast store of letters that Kuhn received following the publication of *Structure*. Many of these are best characterized as fan mail, as they often express enthusiasm for the book. A significant number of these fans are social scientists. Two economists who would later be honored with Sveriges Riksbank Prizes in Economic Sciences in Memory of Alfred Nobel, the so-called "Nobel Prizes in Economics," were among those who contacted Kuhn about *Structure* and its implications for their field, namely, Wassily Leontief and George Stigler. Leontief claims that "what [Kuhn] said about the Natural Sciences certainly applies to Economics too" (see Leontief 1964). Leontief, though, recognizes that Kuhn's book is about the natural sciences. Stigler describes *Structure* as "a splendid book" and compliments Kuhn for arousing him "from his customary lethargy," but questions the extent to which Kuhn's theory of

---

[2] Kuhn would make similar remarks about economics in the late 1980s, in a paper discussing the differences between the natural and the social sciences. Unlike the social sciences, the natural sciences have "paradigms that can support normal, puzzle-solving research" (Kuhn 1991/2000b, 222). Kuhn then notes that his impression is that there are now paradigms capable of supporting normal, puzzle-solving research "in parts of economics and psychology" (Kuhn 1991/2000b, 222–223).

science describes what happens in economics (see Stigler 1963). Specifically, Stigler questions whether Kuhn's view of scientific revolutions matches what occurred with "the marginal utility revolution of the 1870s, [and] the Ricardian revolution" in economics (see Stigler 1963).

Kuhn's reply to Stigler is noteworthy. Kuhn claims that though he knows "nothing about economics" he has "long been one of [Stigler's] distant admirers" (see Kuhn 1963). Consequently, Stigler's "letter meant a great deal to [Kuhn]" (see Kuhn 1963). In fact, Kuhn states, "I am damned grateful to you for writing" (see Kuhn 1963). But Kuhn suggests that the alleged counterexamples Stigler cites from economics do not differ from the examples of theory change that Kuhn is more familiar with in physics. He suggests to Stigler that his alleged counterexamples may in fact qualify as scientific revolutions, provided the changes "demanded some reconstruction of the older theory" because the "proposed innovations were incompatible with parts of the [accepted theories]" (see Kuhn 1963).

The exact conditions that make a change to a scientific theory a paradigm change, that is, a revolutionary change of theory, were far from clear to many readers, and articulating these conditions would prove to be a recurring theme for Kuhn and Kuhnian scholarship. In fact, Kuhn spent much of the later part of his career addressing this issue. Ultimately, he defined changes of theory as lexical changes that violate the no-overlap principle (see Kuhn 1991/2000a, 92–94). For example, during the Copernican Revolution in astronomy, a new lexicon was introduced, one that changed both the extension and the intension of the term "planet." After the revolution in early modern astronomy, the Sun and Moon were no longer regarded as planets, but the Earth was. And planets were no longer characterized as wandering stars, but rather as satellites of the Sun. Social scientists, though, have largely ignored these later developments in Kuhn's view.

The Scientific Status of the Social Sciences

With the publication of *Structure*, many social scientists did exactly what Stigler and Leontief did: reflect on whether Kuhn's theory of scientific knowledge and scientific inquiry were relevant to their own discipline. Indeed, there is a substantial body of scholarly literature by social scientists seeking to determine how well Kuhn's theory of science fits their discipline. It is worth considering a sampling from this vast body of literature.

Eckberg and Hill (1979) suggest that Kuhn's *Structure* fueled a long-standing preoccupation among sociologists with the scientific status of both (i) sociology and (ii) the social sciences in general (see Eckberg and Hill 1979, 933). On the one hand, sociologists tend to compare the social sciences to the natural sciences and regard any differences between the two as indicating that the social sciences are falling short and are not fully scientific (see Eckberg and Hill 1979, 934). On the other hand, given Kuhn's description of the natural sciences, sociologists began to think that the social sciences had as much of a claim to being scientific as the natural sciences had (see Eckberg 1979, 934). That is, given the apparent existence of a sordid side to the research practices that Kuhn describes, some social scientists were less self-conscious about what they had taken to be failings in their own fields.

Political scientists were also inspired to reflect on their discipline. In his Presidential Address to the American Political Science Association (APSA), David Truman (1965) raised doubts about whether political science had ever had a paradigm. Rightly, Truman notes that "a crucial feature of a true paradigm is its precision ... precision in the paradigm permits the investigator to know when something is wrong" (Truman 1965, 866). Political science, though, lacked precise theories and paradigms. Further, Truman pointed out that until recently, individual textbooks in political science had often even lacked consistency in the way they employed theories and background assumptions between chapters (see Truman 1965, 870). A year later, in his Presidential Address to the APSA, Gabriel Almond (1966) argued that political science did have a paradigm, but his description of the discipline leaves one wondering what it is. Both Truman and Almond seem to assume that ideally all political scientists will be united by a single paradigm. Both seem to regard paradigms that are unique to subspecialties within political science as divisive and fractious. This, though, was not Kuhn's view.

Jerone Stephens notes that some political scientists even thought that a paradigm could be imposed on a research community (see Stephens 1973, 482–483; and Walker 2010, 440). In this way, a field like political science could, through some effort, be set on a clear path to being a science. Stephens, though, rightly notes that this was not Kuhn's view, nor is it a practical means for a field to become scientific (see Stephens 1973, 483).

Incidentally, Kuhn explicitly addresses this concern. He suggests that most of the social sciences are proto-sciences, similar in important respects to "fields like chemistry and electricity before the mid-eighteenth century ... [and] the study of heredity and phylogeny before the mid-nineteenth" (Kuhn 1970/2000, 138). As Kuhn explains, "in these fields ... incessant criticism and continual striving for a fresh start are primary forces" (see Kuhn 1970/2000, 138). Consequently, they were not prone to the developmental cycle that Kuhn says characterizes the contemporary natural sciences. And the focus on puzzle solving that is characteristic of fields in the contemporary natural sciences is largely absent.

Kuhn recognizes that social scientists may look at the natural sciences with envy, but he is quite explicit that he is not offering a "therapy to assist the transformation of a proto-science to a science" (Kuhn 1970/2000, 138). Indeed, Kuhn notes that one cannot force a field into the developmental pattern that characterizes the natural sciences (see Kuhn 1970/2000, 139). Even in the natural sciences, where fields have settled into the developmental pattern Kuhn presents in *Structure*, the various fields had to wait for the development of the first paradigm. And not all natural sciences developed a first paradigm at the same time. Notably, the experimental natural sciences did so much later than did the mathematical natural sciences such as astronomy, statics and optics (see Kuhn 1976/1977b, 36). Kuhn refers to the former sciences as the Baconian sciences, which include studies of magnetism and electricity (see Kuhn 1976/1977b, 41–52).

Significantly, Kuhn is not expressing disdain for the social sciences. He is merely noting how they differ from the natural sciences. But this point was largely lost on many, including many social scientists. The economist Mark Blaug (1976), for example, was led to claim that "both Kuhn and Lakatos jeer at modern psychology and sociology as pre-paradigmatic, proto-sciences" (Blaug 1976, 159). Kuhn was not jeering. He was merely trying to advance our understanding of the natural sciences, by noting a significant difference between the natural sciences and the social sciences, the presence or absence of a paradigm.

In fact, quite late in his career, in a relatively short paper, Kuhn took up the issue of what distinguishes the social sciences from the natural sciences. The starting point for his analysis was a disagreement he had had with Charles Taylor, who emphasized the hermeneutical methods employed in the social sciences. Kuhn expressed some skepticism about Taylor's view. Instead, Kuhn suggested that a significant difference between the natural sciences and the social sciences was that the objects studied by the former were more stable. For example, Kuhn noted that through the transition from Greek astronomy to Copernican astronomy "the heavens remained the same" (see Kuhn 1991/2000b, 223). In contrast, the sorts of things that social scientists study, social and political systems, for example, are not like this (see Kuhn 1991/2000b, 223).

Social Scientific Paradigms: Their Ubiquity and Their Relevance

The term "paradigm" or some variation of it is mentioned "478 times in *Structure*, averaging 2.78 times per page" (see Wray 2020, 828). Margaret Masterman, a sympathetic reader of *Structure* and an enthusiastic supporter of Kuhn, famously counted twenty-one different uses of the term in *Structure* (see Masterman 1970/1972). It is not surprising, then, that there has been a lot of confusion about what Kuhn meant by the term "paradigm."

Until around 1970, Kuhn used the term "paradigm" to refer to a variety of different things:

1) sometimes it meant theory, as in the Copernican paradigm or the Newtonian paradigm;
2) sometimes it meant disciplinary matrix, that is, the combination of theory, goals and standards, as in the reigning paradigm in eighteenth-century physics or the reigning paradigm in late-nineteenth-century chemistry; and
3) sometimes it meant exemplar, as in Kepler's mathematical model of the orbit of Mars. (see Wray 2011, chapter 3)

It took Kuhn some time to determine what exactly he meant by "paradigm." It is the latter meaning that Kuhn ultimately settled on. And when he got his thoughts clear on this issue, he preferred to use the term "exemplar."

Despite this ambiguity, the term "paradigm" is now thoroughly integrated into the vocabulary of scientists, both natural scientists and social scientists. Recently, Shiping Tang (2011) has argued that there are "only 11 foundational paradigms in [the] social sciences" (see Tang 2011, 211). Whether or not Tang's claim would stand up to the critical scrutiny of social scientists across the disciplines need not be settled here. The point is that social scientists are now quite comfortable thinking in terms of the paradigms of their disciplines. In this respect, Kuhn has had a wide-ranging impact on the social sciences.

Social scientists, though, are divided on what the correct level of analysis is with respect to finding paradigms in their field. Many sociologists identify what they take to be paradigms of the whole discipline, such as functionalism or conflict theory, theoretical orientations that are used by many sociologists working on different research topics (see Eckberg and Hill 1979, 935). But Eckberg and Hill (1979) claim that this is a mistake. Nothing worth calling a paradigm, they claim, cuts across the whole discipline of sociology. Instead, they believe sociologists should be looking for paradigms specific to much smaller research areas. Thus, we should expect to find paradigms specific to research on the sociology of age and aging, or research on the sociology of political movements.

Two examples from sociology highlight this point. In the 1980s and 1990s sociologists concerned with the study of political stability and instability discussed the new "elite paradigm" (see Field and Higley 1980, and Cammack 1990). The level of analysis in this case is a very specific research topic, that of political stability and instability, and the elite paradigm is presented as an alternative to existing paradigms in the subfield. Similarly, Baltes and Nesselroade (1984) discuss the paradigms operative in the subfield of life-span developmental psychology, an interdisciplinary field that cuts across the fields of sociology and psychology. They note that sociologists and psychologists researching in this area may need different paradigms, as psychologists tend to be more concerned with the biological basis of life-span development than sociologists, who are principally concerned with the social dimensions of that development. The level of specificity in these examples is closer to the sort of thing that Kuhn had in mind when he equated paradigms with exemplars.

George Ritzer (1981) offers another perspective. He argues that sociologists should not follow Kuhn slavishly, but, instead, should refashion the "paradigm" concept to fit the needs of their discipline. Ritzer claims that the notion of a paradigm as an exemplar offers little insight into sociology. He thinks the notion of a paradigm as a disciplinary matrix is more relevant to sociology (see Ritzer 1981, 245).

Incidentally, the language of paradigms has been adopted in some textbooks in sociology. For example, in his textbook *Sociology*, John Marcionis presents students with what he describes as the "three major theoretical paradigms" (see Marcionis 1997, 16–22). These are

1) the Structural-Functional Paradigm associated with Émile Durkheim,
2) the Social-Conflict Paradigm associated with Karl Marx and
3) the Symbolic-Interaction Paradigm associated with Max Weber.

Aside from mentioning Kuhn's name when the term "paradigm" is first introduced, there is no mention of the specifics of Kuhn's view.

Marcionis' textbook reflects what Kuhn claims to have discovered while trying to complete the manuscript of *Structure* at the Center for Advanced Study in the Behavioral Sciences, at Stanford, in the late 1950s. As Kuhn explains, "spending the year in a community composed predominantly of social scientists ... I was struck by the number and extent of the overt disagreements between social scientists about the nature of legitimate scientific problems and methods" (see Kuhn 1962/2012, xlii). In the natural sciences, physics, for example, a student generally does not encounter competing fundamental theories, as they do in sociology, for example. Students of physics, all over the world, are learning the same basic theories. The situation is the same in chemistry, the geosciences and biology. But in the social sciences, a student may learn different competing theories, as there is no consensus in the field on such fundamental issues.

When sociologists discuss paradigms it is now quite common for Kuhn not to be cited at all (see, for example, Cammack 1990, and Field and Higley 1980). This is what Robert K. Merton calls "obliteration by incorporation" (see Merton 1988, 621). A concept becomes so thoroughly integrated into the discourse of a discipline that the originator of the concept is no longer acknowledged, and sometimes even forgotten. Sociologists no longer seem to think it is necessary to alert readers to Kuhn's earlier analysis of the notion of a paradigm, as it is so fully integrated into the research literature. It is now taken for granted that one has a general sense of what a paradigm is.

Let us briefly consider how social scientists other than sociologists reacted to the paradigm concept. Sharer and Ashmore (2003) provide a brief discussion of Kuhn's view and the role of paradigms in archaeology in their textbook *Archaeology: Discovering the Past*. They define a paradigm as "an overall strategy with its unique research methods, theory, and goals" (see Sharer and Ashmore 2003, 109). They suggest that the field of prehistoric archaeology "consists of three research traditions, each of which has evolved under different circumstances, defining its own problems for investigation and the set of data it considers relevant to such problems" (109–110). This clearly sounds like Kuhn's description of a field in a pre-paradigm state. Insofar as prehistoric archaeologists are able to work in different traditions, even attending to different data, there clearly is no agreement about fundamentals.

Economists, too, have raised questions about whether the paradigm or theory is the appropriate unit of analysis for understanding changes in their field. Axel Leijonhufvud (1976), for example, claims that "the 'doctrines' and 'schools' of economics are not all animals of the same species as … Ptolemaic and Copernican astronomy or the Phlogiston and Oxygene [*sic*] theories of combustion; nor … do they very often succeed each other in such clear-cut fashion" (Leijonhufvud 1976, 68). Leijonhufvud claims that often "in economics … several analytical traditions [survive] side-by-side" (1976, 75). Thus, as far as he is concerned, in economics there is no reigning paradigm, as Kuhn suggests there is in the natural sciences. Instead, paradigms or "analytical traditions" are more like tools in a toolbox that can be used separately or together, as the task at hand demands. Alternatively, A. W. Coats (1969) suggests that "economics may ... [have] been dominated throughout its history by a single paradigm – the theory of economic equilibrium via the market mechanism" (Coats 1969, 292).

Political scientists also embraced the notion of paradigms. As noted above, in both 1965 and 1966 the presidents of the APSA drew on Kuhn's *Structure* for insight into the nature of their discipline in their presidential addresses (see Stephens 1973, 476–478). Stephens (1973) provides a comprehensive overview of the scholarly literature in political science that sought to draw insight from Kuhn's *Structure*. He argues that the "paradigm" concept added nothing new to analyses in political science (Stephens 1973, 467). The way political scientists use the term "paradigm," Stephens explains, is more or less equivalent to the way they use the term "theory." Consequently, the paradigm concept sheds no new light on debates in political science about the discipline, its internal dynamics or its research practices. Stephens has a more fundamental criticism of Kuhn's theory of science. He doubts whether Kuhn's account is even an accurate account of the natural sciences (see Stephens 1973, 467).

Stephens also argues that there are important differences in how political scientists and natural scientists are trained. Whereas in the natural sciences, as Kuhn notes, there is a single relatively uniform curriculum taught in a field at any given time, he argues that students in political science are presented with the views of competing schools. In fact, "not only are diverse points between schools presented, but diverse points within schools are presented to students as legitimate ways of studying politics, and they are introduced to this inter- and intraschool diversity through readings of original research reports" (see Stephens 1973, 483). So there are no pretensions that there is a consensus on a single best theory or theoretical framework in political science, unlike the situation in the natural sciences as Kuhn describes it. And in political science students are even made aware of this fact. Again, this would not surprise Kuhn, after his year at the Stanford Center for Advanced Study.

In a 2010 survey article intended to halt appeals to Kuhn and paradigms, Thomas Walker argues that the "paradigm mentality based on normal science and incommensurability has been widely employed, if not internalized, by political scientists" (Walker 2010, 436). But, as Walker rightly notes, the notion of paradigm employed by political scientists is "drawn only loosely from Kuhn's work" (Walker 2010, 433). Walker argues that appeals by political scientists to Kuhn have been unfruitful, and the discipline would be better served by drawing on Karl Popper's writings (see Walker 2010, 434). Popper, after all, explicitly wrote on the social sciences, whereas Kuhn was concerned narrowly with the natural sciences. But, contrary to what Walker seems to imply, Popper had a rather negative view of the social sciences (see Popper 1970/1972, 58). Popper is most explicit about this in *The Poverty of Historicism* (see Popper 1957/1991; and Birner 2018; see also Popper 1970). In this book, Popper argued that many social scientific theories are not capable of generating predictions of the sort that would make them falsifiable, and thus genuinely scientific.

Walker raises two types of concerns with the way Kuhn's work has been used by political scientists, and by scholars in international relations in particular. First, he does not think that a single theory or paradigm dominates in this research subfield in political science. Second, he believes that if political scientists tried to emulate natural scientists as Kuhn characterizes them, it would lead to the suppression of alternative theories and hypotheses (see Walker 2010, 434). Guided by Kuhn, Walker believes that political scientists will be led to "engage in hostile zero-sum turf wars" in their efforts to ensure that their own paradigm is not displaced by an alternative one (see Walker 2010, 434). Walker argues that the "misappropriation of [Kuhn's model] of science ... encourages hyper-specialized tribalism within subfields and furthers the Balkanization of political science as a discipline" (Walker 2010, 434).

This last remark is prescient on Walker's part. Specialization came to play an important role in Kuhn's later philosophy. According to Kuhn, crises are sometimes resolved not only by revolutionary changes of theory, but by the creation of new scientific specialties (see Wray 2011, chapter 7). Kuhn, though, did not see this as a form of tribalism or balkanization, or at least not in a sense that has negative connotations. Instead, he believed that the increasing specialization in the natural sciences explains their great success (see Kuhn 1991/2000a). When a single scientific field or subfield splits into two specializations, scientists can develop conceptual tools more suited to modeling the parts of nature they seek to understand. That is, each of the two new subfields can develop their conceptual resources to suit the research problems that they are concerned with. I have discussed examples of this process elsewhere, including the development of endocrinology from physiology, and the development of virology from bacteriology (see Wray 2011, chapter 7).

These analyses of the paradigm concept by social scientists raise a number of important issues. The following questions are still worth reflecting on:

1) How many paradigms are there in a social scientific field or specialty at any one time? Is a whole field, say, economics, characterized by a single paradigm at any given time?
2) Can competing and incompatible paradigms be used in a social scientific field, and if so, what does this say about the social sciences? And,
3) What is the scope or size of the relevant social group in the social sciences that shares a paradigm? Are they groups of one hundred or so social scientists, or groups of several thousand social scientists?

Indeed, insofar as Kuhn's theory is an apt characterization of the natural sciences, we can ask the same sorts of questions about the natural sciences as well. Many readers, for example, assumed that Kuhn took the relevant social groups to be the members of a whole discipline, like physics. This was a function of the fact that many readers tended to focus on the large revolutions they were most familiar with, like the Copernican Revolution in astronomy, or the Darwinian revolution in the biological sciences. But Kuhn was adamant that the sorts of groups that are affected by a revolutionary change of theory often number around a hundred or fewer members (see Kuhn 1969/2012, 177).

Kuhn's own thinking about paradigms changed as he tried to address criticisms raised by Masterman, Dudley Shapere (1964) and others (see Wray 2011, chapter 3, for a more complete account). Most notably, he thought that Masterman was correct to claim that "a paradigm is what you use when the theory isn't there" (Kuhn 1997/2000, 300). Given this characterization of paradigms, there is no reason to think that the social sciences do not have paradigms. Indeed, it seems far more likely that they do in fact use paradigms to guide them in their research. But even granting that, it does not follow that the social sciences develop according to the developmental cycle that Kuhn says characterizes the natural sciences – that is, starting with a period of normal science that is interrupted by a crisis caused by anomalies, which is then resolved by a scientific revolution that brings in a new theory, followed by a new period of normal science. It is in that framework that puzzle solving plays a crucial role, and that debates about the fundamentals of a field are restricted to periods of crisis.

The Place of Revolutionary Changes of Theory in the Social Sciences

In addition to social scientists' concerns about the relevance of Kuhn's paradigm concept to their fields, there is another question that has occupied social scientists since the publication of Kuhn's *Structure*: Are there revolutionary changes of theory in the social sciences as there are in the natural sciences? Again, this debate has been affected by the ambiguity or lack of clarity in Kuhn's own account of what sorts of changes constitute revolutionary changes of theory. Not surprisingly, many assumed that Kuhn was concerned with wide-ranging scientific revolutions, like the Copernican Revolution in astronomy and the Darwinian revolution in biology. They failed to recognize that he was equally concerned with much smaller scientific revolutions, for example, the revolutions associated with Fresnel's wave theory of light and Dalton's chemical atomic theory (see Wray 2011, 18–19, table 1, for a longer list). Indeed, as we have seen, in the Postscript to *Structure* he makes clear that the sorts of research communities that he had in mind were often constituted by 100 or so scientists (see Kuhn 1969/2012, 177).

Economists have had more to say on the issue of scientific revolutions than other social scientists. Coats (1969) was struck by the fact that revolutionary changes of theory of the sort that Kuhn describes in *Structure* are not at all common in economics. On the one hand, Coats claims that "the most striking example of paradigm-change in economics, the Keynesian revolution of the 1930's [*sic*], possessed many of the characteristics associated with Kuhn's 'scientific revolutions'" (Coats 1969, 293). On the other hand, he argues that "it is now clear that the Keynesian paradigm was not 'incompatible' with its predecessor" (Coats 1969, 293). Coats suggests that there is a reason for this difference between economics and the natural sciences. He suggests that "economic theories … are usually less rigid and compelling than their natural science equivalents" (Coats 1969, 293). Consequently, he believes that we should not be surprised to discover that "the structure of scientific revolutions is much less discernible in economics than in natural sciences" (Coats 1969, 293).

Coats also notes that the practice of "normal science" is different in the social sciences than in the natural sciences. According to Coats,

> economists … have enjoyed considerable success in two phases of the "normal" scientific activity of actualizing the promise inherent in their paradigm, i.e. [1] extending knowledge of "relevant" facts, and [2] improving the "articulation" of the paradigm itself. However, their efforts to [3]

improve the match between the facts and the paradigm's predictions have met with only limited success. (Coats 1969, 292; numerals added)

Thus, though there are some practices in the social sciences that match Kuhn's description of normal science, Coats believes that prediction plays a less significant role in the social sciences. Coats' assessment is a bit misleading. Social scientific theories enable social scientists to predict *patterns*, even if they do not enable them to make precise predictions of singular events, the sorts of predictions that natural scientists are often able to make (see Hayek 1964, 344–345). But this difference may be a consequence of the nature of the objects studied, rather than an indication that social scientific theories are deficient in some way. Friedrich Hayek compared the theories in the social sciences to Darwin's theory of evolution. Both were concerned with complex phenomena. Thus, their inability to deliver precise predictions of single events is not a mark of failure.

Mark Blaug (1976) also argues that disruptive revolutionary changes have played little or no role in economics. Further, he argues that Imre Lakatos' theory of scientific change provides a more accurate description of key changes in economic theory. For example, Blaug claims that "the age-old paradigm of 'economic equilibrium via the market mechanism', which Keynes was supposed to have supplanted, is actually a network of interconnected sub-paradigms; in short it is best described as a Lakatosian SRP [scientific research programme]" (Blaug 1976, 160–161). Further, Blaug claims that "certain puzzles about the Keynesian revolution dissolve when it is viewed through Lakatosian spectacles" (Blaug 1976, 164).

Whether or not Blaug is correct about Lakatos' view and its ability to account for the Keynesian revolution, his assessment of Kuhn is quite unfair. Like so many other critics, Blaug attributes views to Kuhn that he did not accept. For example, Blaug emphasizes Kuhn's suggestion that it was not possible to communicate across theoretical lines, suggesting that incommensurability undermines the possibility for a *rational evaluation* of competing theories. This is precisely the type of reading of Kuhn that Lakatos encouraged, so it is no surprise that Blaug sees more value in Lakatos' account of science (see Lakatos 1970/1972). Blaug also claims Kuhn believed that external factors like personal influence determine which theory is ultimately accepted. This is a caricature of Kuhn's view. Incommensurability, as far as Kuhn was concerned, impedes effective communication across theoretical lines, and makes it challenging to compare and evaluate competing theories. But Kuhn believed that the problem of theory choice is ultimately resolved on the basis of epistemic factors. Kuhn is thus an internalist (see Wray 2011, chapter 9).

Coats was writing in the late 1960s, and Blaug in the mid-1970s. This was before the rise of behavioral economics. One could debate whether behavioral economics constitutes a revolutionary change of theory or not. But, as we saw in the exchange of letters between Kuhn and Leontief and Stigler, arguments can be made on both sides. On the one hand, Herbert Simon saw his work, which replaced the ideal rational agent with "a restricted rational agent," as overturning the accepted theory of mainstream economics (see Sent 2004, § 2 and page 754). On the other hand, Daniel Kahneman's work on bounded rationality is seen as a refinement of, and thus continuous with, the neoclassical approach to economics (see Sent 2004, 749; see also Kahneman 2003). The criteria for determining precisely when a revolutionary change of theory has occurred is far from clear.

Not long after Kuhn's death, the political scientist Nelson Polsby felt it was still worth reflecting on the role of revolutionary changes of theory in his field. Polsby begins by arguing that there is reason to believe that Kuhn's account of revolutions does not provide an accurate characterization of the dynamics of the contemporary *natural* sciences (see Polsby 1998, 205 and 203). He suggests that this should not surprise us, given that Kuhn's *Structure* "was expressly a study of the natural sciences chiefly before 1910" (see Polsby 1998, 199). Nonetheless, Polsby suggests that "for the social sciences ... which have by no means reached a stage of development ... comparable with the hard sciences, Kuhn's account of conflict-ridden revolutions seems quite helpful" (Polsby 1998, 206). He even cites the following three examples from political science that seem to support this claim: "the study of community power ... the economic interpretation of the American constitution ... and the synoptic versus incremental views of administrative behavior" (Polsby 1998, 206).

Anthropologists have also focused on the issue of whether or not there are revolutions in their discipline. Archaeologists in particular have debated whether or not a scientific revolution occurred in their field in the 1960s. Paul Martin, for example, argues that there was a revolution. Until the early 1960s, American archaeologists had aligned themselves and their discipline with history and were concerned primarily with cataloguing and classifying artifacts. During the 1960s, Martin claims, a new paradigm had emerged (Martin 1971, 3). Archaeologists now identified themselves with anthropologists, and thus identified themselves with social scientists. Working within the new paradigm, they now consciously set out to develop a general theory of cultural change, an issue that had not really been on the research agenda of archaeologists before (see Martin 1971, 1–2). Now, like other scientists, they aimed "to establish general laws covering the behavior of the observed events or objects" (Martin 1971, 5).

By the end of the 1970s, some archaeologists and anthropologists were becoming disenchanted with what they saw as attempts to understand their discipline through the lens of the Kuhnian framework. David Meltzer (1979), for example, took issue with the uncritical and reckless application of that framework (Meltzer 1979, 644; also 649). Further, he insisted that there was little evidence to support the popular claim that there had been a revolution in archaeology in the 1960s. Meltzer grants that there were significant *methodological* changes introduced into the discipline. But he rightly notes that methodological changes do not constitute a revolution in Kuhn's sense (see Meltzer 1979, 653). Revolutions essentially involve a change of theory.

There is one interesting feature of the discussion of the relevance of Kuhn's views in archaeology of which it is worth taking note. The debate took place in a national context, specifically in *American* archaeology (see both Martin 1971 and Meltzer 1979). This seems to be one respect in which the social sciences really do differ from the natural sciences. Theoretical orientations are not tied to different nations in the natural sciences. The training and practices of the natural sciences do not differ from country to country. The same theories are accepted in different nations in the natural sciences at any given time. Social scientists, though, are generally more concerned with issues of national concern, which may in fact lead to the acceptance of different theories in different nations in the social sciences (see, for example, Akers 1992). Indeed, this should not be surprising, given that the cause of some type of social phenomenon, petty crime, for example, may be quite different in one society than in another.

The enthusiasm of social scientists for the concept of a scientific revolution may be a function of both the state of the social sciences in the 1960s and 1970s, and the state of the world at that time. In fact, to Kuhn's surprise, his views were frequently debated in the same forums in which Herbert Marcuse's views were being discussed (see Kuhn 1997/2000, 308). This extension of his view struck Kuhn as quite unusual. Though revolutions figure significantly in Kuhn's analysis of scientific change, he insisted that science was a rather conservative institution. In fact, he suggests that what he was trying to make sense of was how "the most rigid of all disciplines, in certain circumstances the most authoritarian, could also be the most creative of novelty" (Kuhn 1997/2000, 308). So "revolution" did not have quite the same connotations for Kuhn as it did for many of his readers.

Kuhn's writing not only affected how social scientists saw their field, but also, at times, it affected them in a more personal way. We can see the impact of Kuhn on the anthropologist William Sewell, Jr. as he became a cultural anthropologist. Sewell's recounting of this experience is expressed in Kuhnian terms, and it happened at the Institute for Advanced Studies, at Princeton, in a seminar in which Kuhn participated. Sewell explains that

> making the cultural turn was ... an exciting but also profoundly troubling step for an adept of the new social history. In my case, and I think in others as well, taking the step amounted to a sort of *conversion experience* – a sudden and exhilarating reshaping of one's intellectual and moral world. In my case, the initial "conversion" took place at the University of Chicago between 1972 and 1974 ... My anthropological turn was powerfully confirmed and deepened during the academic year of 1975–76 at the Institute for Advanced Studies, when I took part in an extraordinary seminar on symbolic anthropology. This seminar, led by Clifford Geertz, included ... [eight other] anthropologists ... [a] sociologist ... and five historians: Robert Darnton, Thomas Kuhn, William Reddy, Ralph Giesey, and myself. (Sewell 2005, 42–43; emphasis added)

In fact, Kuhn mentions Sewell as a person whose company he enjoyed and whose encouragement he appreciated, along with that of Quentin Skinner and Clifford Geertz, while he was at the Institute for Advanced Studies at Princeton in the 1970s (see Kuhn 1997/2000, 320).

In summary, Kuhn has had a pronounced impact on the social sciences. Caught up by his engaging writing, including his many vivid metaphors, many saw in *Structure* things that pleased them or that they perceived to be useful for their own purposes, even if they were not the sorts of things that Kuhn had intended. Many social scientists reflected on the relevance of the Kuhnian concepts of the "paradigm" and "scientific revolution" for their own disciplines, as well as on the extent to which the developmental cycle that Kuhn says characterizes the natural sciences also describes how the social sciences develop. Though no consensus was ever reached, Kuhn's presence was widely felt in the social sciences. And social scientists were in no way put off by the fact that Kuhn had never intended his account of science to be applied to the social sciences.

CHAPTER 6

The Elephant in the Room: The Sociology of Scientific Knowledge

Sociology of science is the specific subfield in the social sciences in which Kuhn has had the greatest impact. In this chapter I want to examine Kuhn's long and complex relationship with the sociology of science, culminating in the emergence of the Strong Programme in the Sociology of Scientific Knowledge (SSK), a group of self-professed Kuhnians. Though the group claimed to be building on Kuhn's view of science, he saw matters quite differently. He did not recognize his own view in their work, and the fact that others associated the Strong Programme with Kuhn's own view created unanticipated and unwelcome problems for Kuhn. The Strong Programme's relativism was often thought to reflect a commitment to relativism on Kuhn's part, and its alleged disregard for rationality was often thought to reflect Kuhn's own disregard for scientific rationality.

I want to begin by tracing Kuhn's early interactions with sociologists of science and examine his own understanding of the relevance of sociology of science to his project. Sociologists of science were interested in Kuhn's work long before the emergence of the Strong Programme. In fact, the generation of sociologists that preceded the Strong Programme were very familiar with Kuhn's research, and drew on it. Further, as we will see, Kuhn believed that sociology was quite relevant to the philosophy of science. His theory of scientific knowledge and scientific change, after all, takes the research community, not the individual scientist, as the relevant unit of analysis. I then examine the emergence of the Strong Programme, its adherents' understanding of their relationship to Kuhn's views, and Kuhn's criticisms of the Strong Programme. As we will see, Kuhn's criticisms of the Strong Programme changed over the years.

Early Engagements with the Sociology of Science

Kuhn's interactions with sociologists of science did not begin in the 1970s with the rise of the Strong Programme. Even before the publication of

103

Structure, Kuhn both interacted with sociologists of science and was identified as a person who might have insights into a sociological study of science.

Writing on behalf of *The Institute for the Unity of Science* in 1952, Philipp Frank asked Kuhn, who was then an associate professor in the General Education program at Harvard, to be part of "a research project under the general title 'sociology of science'" (see Frank 1952). Frank was one of the original members of the Vienna Circle. In fact, he was one of the members of the so-called "First Vienna Circle," the group that had begun meeting even before the First World War (see Frank 1949/1961, 13–14). After leaving Europe in the late 1930s, Frank worked at Harvard, principally as an instructor in physics. But he also taught courses in the philosophy of science (see Frank 1949/1961, 59).[1] Incidentally, though Kuhn drafted a letter in response to Frank, he did not send the letter.

After attending a conference on the history of quantification sponsored by the Social Science Research Council (SSRC), Kuhn was asked in 1959 by the President of the SSRC, Pendleton Herring, for "suggestions with regard to the further development of a sociological approach to the history of science" (Herring 1959). In a reply to Herring, Kuhn claimed that sociological studies of science were crucial to developing a more accurate account of science (see Kuhn 1959b). Kuhn suggested that the SSRC could support a study group to explore the possibilities of "promoting the sociology of science." And he explicitly suggested that "Bob Merton, Bernie Barber, and myself, [and] others . . . could produce [a] . . . concrete proposal" (see Kuhn 1959b).

Incidentally, the paper Kuhn presented at the SSRC conference was a version of his paper "The Function of Measurement in Modern Physical Science" (see White 1963, 84). Kuhn's purpose in that paper was to expose a myth, a myth about the purpose of measurement. The common view, derived from science textbooks, suggests that scientists collect measurements for two reasons: (i) in order to test theories, and (ii) in order to suggest "new scientific laws and theories" (see Kuhn 1961/1977, 182). What Kuhn does in this paper is to compare these two functions "with ordinary scientific practice" (see Kuhn 1961/1977, 183). The focus on scientific

---

[1] The Institute for the Unity of Science would later evolve into the Boston Colloquium for Philosophy of Science (see Isaac 2012, 234). I thank Alisa Bokulich for first alerting me to this connection. George Reisch (2005) argues that though the Institute for the Unity of Science had originally intended to continue the work of the Vienna Circle, Frank's vision for the Institute was at odds with the direction that philosophy of science was taking in America, a direction toward greater professionalization (see Reisch 2005, 294–306).

practice is clearly consonant with a sociological study of science, and it was perceived as such. In a review of the published conference proceedings, the sociologist Harrison White refers to Kuhn's contribution as a "precious gift" (see White 1963, 84). White makes a passing remark that "like any good *sociologist* ... Kuhn is doubtful ... about the wisdom of letting the subjects [of his investigation] know how they actually function" (see White 1963, 84; emphasis added). White parenthetically remarks that "we can claim him as kin in approach if not in training" (White 1963, 84). Incidentally, though White was a sociologist, like Kuhn, he had completed a Ph.D. in physics before finding his professional home in the social sciences (see Breiger 2005, 885–886).

More significantly, Kuhn corresponded with Robert K. Merton in the late 1950s, discussing both his paper on measurement and Merton's influential paper "Priorities in Scientific Discovery" (see Kuhn 1958; Merton 1959; and Kuhn 1959c). Merton was especially struck by Kuhn's discussion of textbook science, a theme that Kuhn would return to in *Structure*, and one that has important sociological implications for understanding the culture of science. Textbook science not only leaves scientists with a distorted sense of the history of their disciplines, but it also supports the view that the growth of science is cumulative, a view Kuhn aggressively sought to undermine. This focus on textbook science is one clear way in which Kuhn was influenced by reading Ludwik Fleck's *Genesis and Development of a Scientific Fact* (see Fleck 1935/1979).[2]

Later, in a letter to Kuhn from 1962, Merton wrote that "more than any other *historian* of science I know, you combine a penetrating sense of scientists at work, of patterns of historical development, and of *sociological* processes in that development" (cited in Cole and Zuckerman 1975, 159; emphasis added). Even though Merton identified Kuhn as a *historian* of science, he found his work to be insightful from a sociological point of view. Merton was not the only sociologist of science with whom Kuhn interacted during this period. He also interacted with Bernard Barber. In fact, as Kuhn notes in a letter to the editor he was working with at the University of Chicago Press, Barber gave Kuhn extensive feedback on a draft manuscript of *Structure* (see Kuhn 1961a). Further, Barber is the *only* sociologist that Kuhn cites in *Structure*.

---

[2] Though Fleck was not a sociologist by training, he was far more interested in the social dimensions of science than contemporary philosophers of science were. Indeed, Fleck was a contemporary of the Vienna Circle.

Kuhn's own views about the sociology of science were quite similar to Merton's (see Kuhn 1977, xxi). Most importantly, like Merton, Kuhn offered a functionalist account of scientific research communities and research practices. That is, he sought to identify those characteristics of scientific research communities or specialties that ensured their success in realizing their goals. The persistence of these features explains the success of science.

Thus, long before the formation of the Strong Programme, Kuhn's work resonated with sociologists of science or those interested in sociological analyses of science.

The perception that *Structure* was relevant to sociologists of science is not a misunderstanding on the part of sociologists. Even though it was a philosophy of science that Kuhn aimed to develop in *Structure*, sociology of science was central to his project. In a letter to Charles Morris, dated July 31, 1953, Kuhn describes the contents of his planned book, tentatively titled *The Structure of Scientific Revolutions*. Kuhn notes that his "basic problem is sociological, since . . . any theory which lasts must be embedded in [a] professional group" (see Kuhn 1953).[3] Kuhn claims that it was Fleck's *Genesis and Development of a Scientific Fact* that "helped [him] to realize that the problems which concerned [him] had a fundamental *sociological* dimension" (see Kuhn 1979, viii; emphasis added; see also Kuhn 1962/2012, xli). And in *Structure*, Kuhn remarks in a footnote that his own work overlaps with Warren Hagstrom's work in the sociology of science. Hagstrom completed a Ph.D. in sociology at the University of California, Berkeley. His influential book, *The Scientific Community*, was published in 1965.

Kuhn's impact on the sociology of science continued to be felt in the 1970s. In an article published in the mid-1970s on the emergence of the sociology of science as a scientific specialty, Cole and Zuckerman (1975), former students of Merton, provide lists of the most-cited authors in the sociology of science between 1950 and 1973. Cole and Zuckerman constructed their rankings on the basis of an examination of publications and citations in nine journals: *American Sociological Review, American Journal of Sociology, Social Forces, Social Problems, Sociology of Education, British Journal of Sociology, Minerva, American Behavioral Scientist* and *Science* (see Cole and Zuckerman 1975, 169, fn. 27). They list the data in five-year periods: Kuhn ranks in the top fifteen of most-cited authors in the period from 1950 to 1954, that is, even before the publication of *Structure*; he ranks

---

[3] I thank George Reisch for drawing my attention to this letter.

in the top ten in the period from 1960 to 1964; and he ranks in the top twenty for the period from 1965 to 1969. And in the four-year period from 1970 to 1973, the final period for which data are reported, Kuhn ranks again in the top ten. When Cole and Zuckerman wrote this paper, they noted that "there are no signs that … Kuhn [is] … turning away from [his] interest in the sociology of science" (1975, 154). Thus, in the mid-1970s Kuhn was identified as an important sociologist of science *by* sociologists of science.

Even after the publication of *Structure*, Kuhn continued to describe his own approach to the study of science as "quasi-sociological" (Kuhn 1997/ 2000, 310). In the Preface to *The Essential Tension* Kuhn claims that his own work is "deeply sociological, but not in a way that permits that subject to be separated from epistemology" (1977, xx).[4] Specifically, he claims that "scientific knowledge is intrinsically a *group* product" (xx; emphasis in original). Kuhn elaborated on these themes when he received the J. D. Bernal Award from the Society for the Social Studies of Science (4S). Kuhn claimed that

> *Structure* is sociological in that it [1] emphasizes the existence of scientific communities, [2] insists that they be viewed as the producers of a special product, scientific knowledge, and [3] suggests that the nature of that product can be understood in terms of what is special in the training and values of those groups. (Kuhn 1983a, 28; numerals added)

In the talk he gave on the occasion of that award, Kuhn confessed that when he wrote *Structure,* he knew very little about sociology. In fact, he claimed that "[he] proceeded to make up the sociology of such communities as [he] went along … [drawing] it from [his] experience with the interpretation of scientific texts supplemented by [his] experience as a student of physics" (1983, 28). Kuhn claims that his guiding question was: "*why [has] the special nature of group practices in the sciences been so strikingly successful in resolving the problems scientists choose*"? (1983, 28; emphasis added). It is worth remembering that the sociology of science was in its infancy in 1962 when Kuhn initially published *Structure* (see Cole and Zuckerman 1975). So, there was far less of it to draw on at that time. Indeed, even as late as 1968, Kuhn described the sociology of science as underdeveloped (Kuhn 1968/1977, 121). Consequently, it is not surprising that Kuhn felt compelled to "make up" the sociology as he went along.

---

[4] Incidentally, Hans Reichenbach had a similar view about the relevance of sociology to epistemology. He claimed that "the first task of epistemology [is its] *descriptive task* … Epistemology in this respect is part of sociology" (1938/2006, 3).

A more sympathetic way of saying it is that he made some theoretical conjectures, based on what he had experienced in his own training as a physicist and on his interpretation of scientific texts.

A New Beginning in the Sociology of Science

With the formation of the Strong Programme, Kuhn's influence on the sociology of science became even more pronounced. But ultimately Kuhn would be dismayed at the direction in which the sociology of science would develop, even though its development owed something to the publication of *Structure* (see Kuhn 1977, xxi; and 1997/2000, 316–317).

Largely based in Britain, the Strong Programme in the SSK developed both (i) in response to Kuhn's *Structure*, and (ii) in reaction to Merton's approach to the sociology of science. Merton and his students worked within a functionalist framework, and often conducted quantitative studies of science. The Strong Programme, on the other hand, often conducted qualitative studies, including historical studies of scientific controversies and ethnographic studies of specific scientific research settings. They were especially intrigued by what Kuhn called "normal science," in contrast to philosophers of science, who were far more intrigued by or preoccupied with revolutionary changes of theory.[5] Kuhn's impact was so great that the proponents of the Strong Programme identified themselves as Kuhnians of sorts, despite Kuhn's own misgivings.

Kuhn's relationship with the new sociology was not always strained. In fact, he was quite supportive at first. He even served on the editorial board for *Science Studies*, a key publishing venue for the proponents of the Strong Programme, from the journal's inception in 1971 until 1976, when it was renamed *Social Studies of Science* (see Kuhn 1976). The editor, David Edge, initially asked Kuhn to serve on the editorial board because he anticipated that contributing authors, especially sociologists of science, would draw on Kuhn's theoretical framework (see Edge 1969). And, indeed, many contributing authors were inspired by Kuhn's work. In fact, in the first three years, over 26 percent (thirteen of forty-nine) of the articles, discussion pieces and notes cited Kuhn, and most often Kuhn's *Structure*.[6] In 1976, when Kuhn insisted that he should no longer serve on the editorial board

---

[5] Philosophers did not totally neglect "normal science." Popper attacked the very notion of "normal science" at the Bedford College conference in 1965 (see Popper 1970/1972). I will return to this issue in Chapter 9.

[6] In contrast, about 14 percent (seven of forty-nine) of the articles, discussion pieces and notes cited Merton.

for the journal, he noted that he "liked the journal from the start, and [he] thinks it has been improving steadily and serving an increasingly significant function as it does so" (see Kuhn 1976). Thus, initially, Kuhn was quite supportive of and enthusiastic about the new research in the sociology of science.

Structure also played a key role in the curriculum for the Science Studies program at the University of Edinburgh, one of the principal training grounds for the Strong Programme. *Structure* was the central text in a course titled "A Philosophical Approach to Science" (Bloor 1975, 507). In fact, as early as 1967, David Edge was "teaching a course based on 'The Structure of Scientific Revolutions'" at Edinburgh (see Edge 1967).[7] No wonder the proponents of the Strong Programme identified themselves as Kuhnians. Their training and initiation required familiarity with Kuhn's *Structure*.

Though clearly influenced by Kuhn, the Strong Programme took liberties with his work. Attempting to draw out the logical consequences of his view, they ended up developing a position that Kuhn did not recognize as his own.

In fact, the proponents of the Strong Programme are correct to insist that much of what Kuhn says in *Structure* is relevant to sociologists of science. For example, Kuhn is concerned with the social structure of scientific research communities, the units that create scientific knowledge and validate knowledge claims. Further, he is concerned with understanding the various *social changes* that a research community undergoes in its efforts to address the challenges encountered while pursuing its research goals. In crisis, for example, a specialty is less beholden to a single, relatively inflexible interpretation of the accepted theory. Instead, scientists work with multiple, even somewhat incompatible, interpretations of that theory. They even consider alternative theories, something that is not a real option during phases of normal science. The norms of the community are thus significantly relaxed during a crisis period. These social changes in the research community, Kuhn claimed, are integral to understanding why science is so successful.

Kuhn also draws attention to the important role that socialization plays in scientific training, including the processes by which scientists-in-training learn to perceive, which essentially involves learning to attend to

---

[7] Brian Easlea reports that in 1968–1969, Kuhn was an essential part of the curriculum for an introductory course in history and social studies of science at the University of Sussex. The required reading included Kuhn's paper "The Function of Dogma in Scientific Research," but *Structure* was included in a list of further readings (see Easlea 1973).

the relevant similarities and differences in the phenomena. Importantly, the notion of "relevant similarities" is relative to a theory or conceptual scheme, and thus relative to the research community. And Kuhn's remarks on textbook science are clearly relevant from a sociological point of view. He suggests, for example, that textbooks provide brief distorted histories that support the picture of scientific knowledge as strictly cumulative, a picture that serves the interests of scientists working on the sorts of esoteric puzzles that characterize work at the research frontier.

All of these aspects of Kuhn's account of science attracted the attention of the proponents of the Strong Programme. But the aspect of Kuhn's account that had the greatest influence on them was, without a doubt, Kuhn's theory of concept learning and concept application (see Kuhn 1974/1977, 293–319). In fact, Kuhn's example of the child learning various bird concepts from his father is discussed at length both by Barry Barnes, in *T. S. Kuhn and Social Science* (see Barnes 1982), and Barnes, Bloor and Henry, in *Scientific Knowledge: A Sociological Analysis* (see 1996, 49–53).

Kuhn's account of concept application is the foundation of the Strong Programme's finitism (see Barnes et al. 1996, chapter 3). According to the Strong Programme's finitism, every act of classification is underdetermined by logic and evidence (see Barnes et al. 1996, 27–35, and 54). As a result, at every instance decisions must be made about how to apply concepts, and what is to count as an instance of the class or kind designated by a particular concept. This is an extreme form of nominalism. Significantly, for the Strong Programme's agenda, the open-endedness of concept application makes room for the influence of social factors, like interests and relations of power. In doing so, the Strong Programme's adaptation of Kuhn's account of concept application provided a rationale and legitimation for sociological studies of the *content* of scientific theories. This was a significant departure from the sociology of science in the Mertonian tradition, where the principal focus was on understanding the institutions and norms that were constitutive of science. Further, this was a significant departure from Kuhn's own view. Kuhn, after all, claimed that because only the experts in a field are capable of evaluating the merits of competing theories in the natural sciences, theory choice is shielded from the influence of many social factors.

Further, strictly speaking, the Strong Programme's finitism should not be identified with Kuhn's own view of concept application. Even Barnes recognized this, for he describes finitism as *based* on a particular reading of Kuhn's view (see Barnes et al. 1996, 207, fn. 1; Wray 2011, chapter 9). Barnes even hints that finitism may be a more extreme position than

Kuhn's own account of concept application. Though Barnes says that he believes Kuhn is a finitist, he also insists that the commitment of the proponents of the Strong Programme to finitism is not in any way dependent on Kuhn endorsing finitism (see Barnes 1982, 34–35). Thus, it should not surprise us to find that Kuhn did not regard the Strong Programme's view as identical to his own.

I have argued elsewhere that finitism is a far more radical form of nominalism than that which Kuhn accepted. Whereas the proponents of the Strong Programme regard *every* act of classification as underdetermined, Kuhn merely insisted that there is no unique way to classify things in the world (see Wray 2011, Chapter 9). From Kuhn's perspective, the proponents of the Strong Programme exaggerate the open-endedness of concept application. As far as Kuhn was concerned, in periods of normal science concept application is generally unproblematic. Working in a normal scientific tradition, scientists generally take for granted that the concepts supplied by the accepted theory are adequate for representing the structure of the world, and phenomena are most often readily classified on the basis of the accepted theory. When problems do arise, a scientist's first impulse, generally, is to assume that they have done something wrong. That is, scientists do not initially assume that the theory they are working with is flawed, or inadequate for the task at hand. Their early training has conditioned them to see the world through the lens of the accepted theory, unreflectively fitting nature into the conceptual boxes supplied by that theory.

It is only when persistent anomalies are encountered that problems arise. Indeed, for Kuhn, an anomaly just *is* something that cannot be classified given the conceptual resources supplied by the accepted theory. The existence of anomalies, though, as far as Kuhn is concerned, is not a problem. They are the sources of research puzzles, the sorts of things that engage scientists in the everyday work of normal science. It is really only when scientists try to resolve persistent anomalies that the underdetermination of concept application becomes important. In these situations scientists must often take an active role in determining how a particular concept is to be applied. Indeed, in such situations disputes often arise over what is the proper scope of a concept.

The proponents of the Strong Programme, on the other hand, believe that concept application is not so constrained even in periods of normal science. Thus, the everyday research practices of science are as problematic from a sociological point of view as are revolutionary changes of theory. In both situations, the scientist must make decisions about how they are to apply concepts, even if in most normal scientific research situations they tend to unreflectively apply concepts in an unproblematic way. The fact that scientists are able to do certain activities unreflectively should not lead them to be complacent, and think that in those cases there are no social causes involved. This is one of the principal differences between Kuhn's own view and that of the Strong Programme.

Kuhn's Criticisms

Not surprisingly, many were alarmed by the way in which the Strong Programme were interpreting Kuhn's work. For example, in a letter to Kuhn, dated 1976, Merton expressed concern about the "Kuhnians" and an alleged "Kuhn-vs-Merton" dispute that Merton believed the self-proclaimed Kuhnians were manufacturing (see Merton 1976).

Kuhn was also beginning to be concerned. Kuhn's unease with the work of the Strong Programme changed over time, influenced both by his own deepening understanding of *their* project, and in response to the evolution of the Strong Programme. It is worth distinguishing three separate criticisms Kuhn developed against the Strong Programme: one presented in the 1970s, one in the 1980s and one in the 1990s. It is really only the latter criticism, though, that gets at something highly significant.

Initially, in the mid-to-late 1970s Kuhn was concerned that the proponents of the Strong Programme failed to understand the role that values play in science. He contrasted their view with Merton's view, a view that he regarded as essentially correct. Whereas Merton believed that science was characterized by a set of values which are more or less stable throughout the history of modern science, the proponents of the Strong Programme denied this. On Kuhn's reading, the proponents of the Strong Programme believed that "values vary from community to community and from time to time" (see Kuhn 1977, xxi). This is evident from the Strong Programme's detailed historical studies, which emphasize the *contingencies* that influence the resolution of disputes in science. These studies purport to show that there is no set of values constitutive of *all* science. Consequently, as Kuhn puts it, the proponents of the Strong Programme "think it absurd to conceive the analysis of values as a significant means of illuminating scientific behavior" (Kuhn 1977, xxi).[8] Kuhn objected to this view, insisting that the shared values constitutive of science play an

---

[8] In a footnote, Kuhn suggests that this criticism "has surfaced frequently . . ., especially in the journal *Social Studies of Science*" (Kuhn 1977, xxi, fn. 9).

important role in enabling scientists to realize their epistemic goals (1977, xxi–xxii).

Around this same time, in an effort to defend himself from the charge that the view presented in *Structure* undermines the objectivity of science, Kuhn argued that the following five values have played an important role throughout the history of science, whenever scientists have evaluated competing theories: accuracy, consistency, breadth of scope, simplicity and fruitfulness (see Kuhn 1973/1977, 321–322). Kuhn contends that science is objective in virtue of the fact that scientists appeal to the *same* values in their decision-making. These are *shared* values, and he contrasts them with individual subjective values, values that are a function of an individual's training and personal history (see Kuhn 1973/1977, 337–338). But even these values, Kuhn claimed, can play a constructive role in science, providing a source of novel ideas when a field is in crisis.

This criticism of the Strong Programme seems not to have generated any traction, as it has been largely ignored by both sociologists of science and philosophers of science.

By the early 1980s, when Kuhn was being honored by the Society for the Social Studies of Science, he had a different concern with the Strong Programme. He was concerned that its studies of science focused on the wrong sorts of values and interests. As far as Kuhn was concerned, these sociological studies focused "pre-dominantly [on] socio-economic interests," and mistakenly "excluded the special cognitive interests inculcated by scientific training," like "love of truth … [and] fascination with puzzle solving" (Kuhn 1983a, 30). Kuhn thus felt that proper sociological studies of the content of science must attend to the specific interests that motivate scientists, that is, their cognitive interests.

Let me highlight the difference between this concern and Kuhn's earlier concern. The earlier concern was that the proponents of the Strong Programme discounted the significance of any values in explaining scientists' behavior. Kuhn's more recent concern is framed in terms of the interests that motivate scientists. Here Kuhn claims that the proponents of the Strong Programme are preoccupied with the wrong sorts of interests, and wholly neglect a class of interests that are indispensable to understanding the success of science. Science, after all, is a cognitive activity, and has epistemic aims. An adequate account of science must take account of these considerations.

To some extent this is not a fair assessment of the Strong Programme. Though the proponents of the Strong Programme emphasize the role of interests and negotiation in determining the outcome of scientific disputes,

Barnes is quite explicit that scientists' interests include the "interest in prediction and control" (see Barnes 1982, 110; see also Hesse 1982a, 329). Indeed, this seems to be lost on most philosophers of science, who tend to treat the Strong Programme with great skepticism, if not contempt. But the proponents of the Strong Programme clearly recognize that scientists are concerned with prediction and control.

Later, by the 1990s, Kuhn had raised a third criticism against the Strong Programme. He complained that its studies of science "leave out the role of [nature]. Some of these people simply claim that it doesn't have any [role], that nobody has shown that it makes a difference" (Kuhn 1997/2000, 317). Though Kuhn acknowledged that by the 1990s the proponents of the Strong Programme no longer held this extreme view, he believed they had failed to clarify what role nature plays in resolving disputes in science (Kuhn 1997/2000, 317).

Kuhn reiterates this criticism in a letter, from 1993, to the historian of science Charles Gillispie. Gillispie initiated the exchange, responding to Kuhn's Rothschild Lecture, "The Trouble with the Historical Philosophy of Science," in which Kuhn discusses the Strong Programme. Gillispie remarks that "it won't surprise you that I entirely share your views about the strong program, which has seemed to me ... a candidate to exemplify Descartes's dictum about there being no position so absurd that it has not been held by philosophers" (Gillispie 1993).

Kuhn, though, was less dismissive of the Strong Programme than Gillispie was. In fact, Kuhn takes some responsibility for the development of the Strong Programme, or at least acknowledges that he played "a causal role in their ... development" (Kuhn 1993b). Kuhn suggests the Strong Programme was developed in response to a false dilemma. Specifically, he claims that it is a response to the following dilemma: "either [1] there is a mind-independent reality which scientific research comes closer and closer to getting right or else [2] there's no role for rational discussion of solid experimental results in a proper understanding of science" (Kuhn 1993b; numerals added). In response to Gillispie, Kuhn claims that "it's that dichotomy that has let in the strong program" (Kuhn 1993b). However, he insists that it is a false dichotomy. He maintains that the ability of science "to solve more and more refined puzzles" is a consequence of scientists responding to experimental results, even though scientists are not converging on some goal set by nature in advance (see Kuhn 1993b). Thus, he insists that his own position is a third alternative.

There is some basis for Kuhn's third criticism of the Strong Programme. They do suggest, at times, that nature plays a negligible role in resolving disputes in science. For example, in an explanation of how concepts are applied, Barnes claims that "*nature* sets *no constraints* on the form of the routine which is produced" (Barnes 1982, 29; emphasis added). Elaborating, Barnes explains that "*any way* of developing the accepted usage of a concept could equally well be agreed upon, since any application of a concept to an instance can be made out as correct and justified by the invocation of an appropriate weighting of similarity against difference" (Barnes 1982, 29; emphasis added).

These remarks bring into focus the radical nature of the Strong Programme's finitism. The world, or whatever we might call whatever it is that scientists study, is extremely malleable. But significantly, Barnes still emphasizes the fact that an *individual* scientist is not at liberty to do whatever they please, that is, not if they hope to influence the development of science. Central to the Strong Programme's view is the agreement among scientists. Innovative applications of concepts must be picked up by other scientists if they are to have an impact on the development of scientific knowledge.

Kuhn definitely did not believe that what scientists study is infinitely malleable. He insisted that nature plays a significant role in shaping scientists' beliefs. Nature places real and firm constraints on what we can believe. In Kuhn's words, the world is "not in the least respectful of an observer's wishes and desires; quite capable of providing decisive evidence against invented hypotheses which fail to match its behavior" (Kuhn 1991/2000a, 101). Kuhn thus believed that an adequate understanding of science requires attention to the role nature plays in resolving disputes. It is in this respect that Kuhn thought of his view as fundamentally different from the view of the Strong Programme.

Interestingly, Mary Hesse had alerted Kuhn to a concern she had with *Structure*, a concern that is very similar to Kuhn's about the role of nature in the Strong Programme's studies. As Kuhn explains, Hesse wrote "a very favorable review" of *Structure* when it was published, but when they met shortly afterwards she said: "'Tom, the one problem is now you've got to say in what sense science is empirical' – or what differences observation makes" (Kuhn 1997/2000, 286).[9] Kuhn was taken aback by Hesse's remark. He explains that "I practically fell over; of course she was right but I wasn't seeing it that way" (Kuhn 1997/2000, 286). Incidentally, Hesse had trained

---

[9] Kuhn had first met Mary Hesse when he went to England during the final summer of his fellowship with the Harvard Society of Fellows (see Kuhn 1997/2000, 285).

both David Bloor, one of the early and most influential proponents of the Strong Programme, and Steve Fuller, a pioneer in social epistemology and critic of Kuhn (see Hallberg 2017, 168).

It seems that the claim that Kuhn is the father of the Strong Programme has some credibility, given Hesse's remarks about the need for Kuhn to explain in what sense science is empirical. Their work really is a product or extension of Kuhn's *Structure*, even if it does depart from his views in significant ways. And the similarities between their views and his make them vulnerable to some of the same criticisms. Thus, even in the twenty-first century Michael Friedman could discuss the Strong Programme under the provocative title of "the Post-Kuhnian Predicament" (see Friedman 2001, 48).

Despite Kuhn's responses to the Strong Programme, many philosophers have seen and continue to see significant parallels between his view and the Strong Programme's view of science. Kuhn's remarks at the end of *Structure* about how appeals to truth explain very little in science *may* sound like an anticipation of the Strong Programme's symmetry principle (see Kuhn 1962/2012, 169–172). But this was not Kuhn's intention. The Strong Programme's symmetry principle is a methodological principle instructing sociologists of science to seek explanations for both true and false beliefs in terms of social causes (Barnes and Bloor 1982). In invoking this principle, the proponents of the Strong Programme want us to resist the temptation of thinking that true beliefs are caused by our successful interaction with nature, and only false beliefs have social causes. The symmetry principle is thus a corrective to reducing the sociology of science to the sociology of error. This has been a common view about the proper scope of the sociology of science among philosophers of science. For example, Larry Laudan has argued that "*the sociology of knowledge may step in to explain beliefs if and only if those beliefs cannot be explained in terms of their rational merit*" (Laudan 1977, 202; emphasis in original). But the proponents of the Strong Programme have insisted that even true beliefs have social causes. Further, and more importantly, they have insisted that social causes of belief are not necessarily distorting. That is, when social factors influence belief, we should not necessarily regard them as contaminants, inevitably frustrating scientists in their pursuit of the truth. This last point is especially important, and it has been largely lost on philosophers of science.

Like the proponents of the Strong Programme, Kuhn believed that scientists are influenced by a variety of social factors. And many social factors that influence scientists can play a constructive role in their pursuit of their cognitive goals. But, unlike the proponents of the Strong Programme, Kuhn was a committed *internalist*. He did not believe that the outcomes of scientific disputes are determined by factors external to science, at least not in the natural sciences. Kuhn did, however, acknowledge that when the available evidence does not unequivocally support one of the competing hypotheses, scientists are influenced in their decision-making by various subjective factors (see Kuhn 1973/1977). For example, Kepler's neo-Platonism, his preoccupation with finding a deep mathematical structure in the cosmos, led him to accept the Copernican theory more than a decade before Galileo's telescopic evidence was gathered (Kuhn 1973/1977, 323). This helped ensure that the Copernican theory was kept alive. According to Kuhn, the influence of such subjective factors ensures that there is an effective division of labor in the research community, and that the viable competing hypotheses are developed. But Kuhn believed that when science is working well, controversies in science are resolved on the basis of a consideration of the epistemic merits of the competing views (see Wray 2011, chapter 9).

It is Kuhn's internalism that both distinguishes his view from that of the Strong Programme and makes his account a contribution to the *epistemology* of science, rather than to the sociology of science. Disputes in science are ultimately resolved on the basis of a consideration of (i) the available evidence and (ii) the various theoretical values. The proponents of the Strong Programme, on the other hand, reject the internalism/externalism distinction. Their point is that the sorts of factors that philosophers typically regard as external can play an integral role in determining the outcome of scientific disputes (see Pinch 1979; Barnes 1982, 118). And, as far as they are concerned, nothing is gained in trying to draw such a distinction.

Indeed, Kuhn's *Structure* is both sociological and epistemological. As Paul Hoyningen-Huene put it, Kuhn believed that "philosophy and sociology of science cannot be practiced independently of each other" (Hoyningen-Huene 1992, 491).[10] Ultimately, though, Kuhn's *aims* were epistemological. That is, he was aiming to make a contribution to the philosophy of science. But he believed that an adequate philosophy of science will have a sociological basis. That is, he believed that an

---

[10] Hoyningen-Huene, for example, argues that "the treatment of *some* allegedly pure philosophical questions, such as those about the dynamics of theories . . . necessarily involves sociological aspects" (Hoyningen-Huene 1992, 491–492).

understanding of the social dimensions of inquiry and the structure and dynamics of scientific research communities was indispensable to developing an adequate understanding of how and why scientists succeed in achieving their epistemic goals.[11]

My aim in this chapter has been to examine Kuhn's complex relationship with the sociology of science, and especially the Strong Programme in the Sociology of Scientific Knowledge. Though Kuhn conceived of his project as sociological in some important respects, the Strong Programme developed in directions that were inimical to Kuhn's understanding of science. Most significantly, he felt they neglected to account for the influence of scientists' epistemic interests, and they failed to account for how nature constrains scientists.

I believe that the proponents of the Strong Programme did contribute to our understanding and appreciation of Kuhn's views, as well as our understanding of science. Most significantly, I believe the proponents of the Strong Programme are responsible for drawing attention to the notion of normal science, something that philosophers of science have only lately begun to study with care. Their studies have contributed to making the everyday practices of science, that is, normal science, something worthy of the attention of philosophers of science. Indeed, the philosophy of science in practice movement attests to this. Further, as I have argued elsewhere, philosophers of science could benefit from either working with sociologists of science or drawing on sociological studies of science, especially as they try to understand how the social dimensions of science contribute to its success (see Wray 2011). This latter issue is now widely recognized as an important part of the epistemology of science.

---

[11] I am not alone in insisting that Kuhn's project is an epistemology of science. Robert Nola (2000) goes to great lengths to argue (i) that the Strong Programme misappropriated Kuhn's work, and (ii) that Kuhn's concerns were continuous with the concerns of philosophers of science (see Nola 2000, 89). Nola characterizes Kuhn's view in the following way: "Even though sociology of *science* can play a role in the individuation of scientific communities, for Kuhn sociology of *scientific knowledge* plays very little role in theory choice and none in his account of the justification of his principles of theory choice" (Nola 2000, 80). Nola, though, believes that Kuhn's project ultimately fails. I think Nola underestimates the role that sociology plays in Kuhn's philosophy of science.

Kuhn and the History of Science

In Part III, I explore Kuhn's complex relationship to the history of science. On the one hand, he was clearly a historian of science, for much of his career was spent working in history departments and teaching courses in the history of science. On the other hand, Kuhn's research has not left as significant an impact on the history of science as it has left on the sociology of science or the philosophy of science. Further, contemporary historians have a rather dismal assessment of *Structure* as a contribution to the history of science. I argue that this assessment of *Structure* is really irrelevant, as *Structure* was not intended to be a contribution to the history of science. It was intended as a contribution to the philosophy of science.

CHAPTER 7

Copenhagen, 1962–1963: A Return to the History of Science

The year 1962 was a very significant one for Kuhn. It was in 1962 that *Structure* was finally published. Kuhn was finally able to complete the project that he had conceived in embryonic form in 1947, when he had his Aristotle experience. Also, Kuhn spent the 1962–1963 academic year in Copenhagen, working on the History of Quantum Physics Project, a project sponsored by the American Physical Society (see Kuhn 1997/ 2000, 302). John Van Vleck, Kuhn's Ph.D. supervisor, was one of the people who asked him to lead the project (see Kuhn 1997/2000, 302–303). With this undertaking, Kuhn was returning to both the history of science and physics.

The timing of this project was quite fortunate, as Kuhn had just had a significant career setback at Berkeley, and he probably welcomed the opportunity to be away from Berkeley for a while (see Kuhn 1997/2000, 301–302). Kuhn had been denied a promotion to full professor in the Department of Philosophy, and as a result he was now quite discontent at Berkeley. As Kuhn reports, "I was extraordinarily angry ... and very deeply hurt, I mean that's a hurt that has never altogether gone away" (Kuhn 1997/2000, 302).

Kuhn's appointment at Berkeley was somewhat unusual from the start, which may explain why things turned out as they did. He was initially offered a job in the Philosophy Department, but during the hiring process, he was offered an appointment in the History Department as well.[1] Thus, his appointment was in *both* the Philosophy Department and the History

---

[1] Reflecting on his hiring at Berkeley later in his life, Kuhn notes "the philosophers at Berkeley wanted to hire a historian of science. They didn't know that they didn't want one, they didn't know that this was not a philosophical discipline" (Kuhn 1997/2000, 294). The enthusiasm at this time for *history and philosophy of science* programs and centers in America was widespread. Margaret Rossiter provides a useful history of the development of *history of science* Ph.D. programs in the United States, from 1954 to 1966, noting the importance of the National Science Foundation's *History and Philosophy of Science Program* in supporting this development (see Rossiter 1984).

Department, and he taught courses in both disciplines (see Kuhn 1997/ 2000, 294; see also Dupree and Kuhn 1959, 172). At the time of his hiring, Kuhn was delighted to be offered an opportunity to work in a Philosophy Department. As he explained, "I jumped at the change, because I wanted to do philosophy" (Kuhn 1997/2000, 294). When he took the job at Berkeley, he was still busily working on the manuscript of *Structure*, which he conceived as a contribution to the philosophy of science. So his being in a philosophy department was as an especially congenial arrangement for Kuhn.

Kuhn's disappointment with the Philosophy Department at Berkeley was precipitated by an unsolicited job offer he received from Johns Hopkins University, an offer of a job at the rank of full professor. It was a very attractive offer, involving hiring of junior colleagues as well (see Kuhn 1997/2000, 301). Kuhn notes that when he returned from visiting Johns Hopkins, he was asked by the administration at Berkeley "what it would take to make [him] stay at [Berkeley]" (Kuhn 1997/2000, 301). Kuhn said that he wanted a promotion to full professor. He expected, it seems, to be promoted to the rank of full professor in both departments, Philosophy and History. This is what ultimately led to the incident with the Philosophy Department.

Edward Strong, who was then the Chancellor at U. C. Berkeley, and was involved with this case, provides the following recollections:

> Tom Kuhn came to me when I was Chancellor and said that he had had this very attractive offer to go to Johns Hopkins, that he would prefer to remain at Berkeley if he could receive a *full-time appointment* in the Department of Philosophy. I said I was certainly in favor of that, but of course that required the approval of the Department of Philosophy. So I conferred with the chairman and asked him to bring the matter before the Department of Philosophy ... I think, unhappily for Berkeley, because of what Kuhn represented to Berkeley in the field of the history of science, the philosophy faculty decided that they didn't want the half-time appointment to be increased to a full-time appointment. (Strong 1992, 285; emphasis added)

Strong's recollections suggest that there may have been a misunderstanding between what he thought Kuhn wanted and what Kuhn in fact thought he had asked for. Kuhn wanted to be promoted to full professor in philosophy (as well as in history) (see Kuhn 1997/2000, 301). But Strong's recollection seems to be that Kuhn wanted his position increased to full-time in the Philosophy Department. What exactly the Philosophy Department voted on, we do not know. But the result was quite consequential for the direction Kuhn's career subsequently took.

Kuhn was denied a promotion to full professor in the Philosophy Department, but granted a promotion to full professor in the History Department. Oddly, Strong mistakenly recalls that *Kuhn took the job* at Johns Hopkins. He reports:

> Tom went to Johns Hopkins, and not so long after, he published a work that attracted international attention and gained him an international reputation as a leading historian and philosopher of science. We could have kept Tom at Berkeley with cooperation from the department. I think it was the department's loss; I know it was Berkeley's loss. (Strong 1992, 285; emphasis added)

In fact, Kuhn went to Denmark in the 1962–1963 academic year, and then returned to teach one more year at Berkeley. The year in Copenhagen was a timely reprieve from this debacle. It was while Kuhn was in Copenhagen that he was offered a position at Princeton, which he did take, starting in the fall of 1964 (see Kuhn 1997/2000, 302).

This was not Kuhn's first significant professional setback. Earlier, in the mid-1950s, he had been denied tenure at Harvard. That had been a very stressful experience for him. Kuhn explains that "[he] was one of those people who was at least in real danger of breaking up because Harvard didn't want them there. That was something that happened to people who'd spent too much time around Harvard" (Kuhn 1997/2000, 289). No doubt, the experience with the philosophers at Berkeley would have stirred up these old feelings.

A Return to History of Science

The History of Quantum Physics Project would bring Kuhn back to historical scholarship for a while. Principally, the project involved constructing an archive based on interviews with the still-living participants of the revolution in early twentieth-century physics. Copenhagen was an ideal location for the project. Niels Bohr was still alive. And, as a consequence, many physicists involved in the revolution were willing to travel to Copenhagen to be interviewed. The project was supported generously by funding from the National Science Foundation's History and Philosophy of Science Program. In fact, the History of Quantum Physics Project was one of the two most costly projects supported by the NSF's HPS Program (see Rossiter 1984, 102).[2] The other major project supported by the

---

[2] Margaret Rossiter reports that the American Institute of Physics was awarded "$203,000 in fiscal year 1961 and $23,600 three years later … for the Archives for the History of Quantum Physics Project (naming John A. Wheeler as principal investigator), distributed over 1962–1966" (Rossiter 1984, 102).

Program was "the *Dictionary of Scientific Biography*," with a grant to the American Council of Learned Societies (Rossiter 1984, 102).

In a number of respects, this was a fruitful time for Kuhn. Together, with John Heilbron, Paul Forman and Lini Allen, Kuhn published *Sources for History of Quantum Physics* (see Kuhn et al. 1967). Heilbron and Forman had been Kuhn's students, and Allen was an assistant to the project, the principal aim of which was to produce the *Sources for History of Quantum Physics*. In addition, though, Kuhn wrote a lengthy paper with Heilbron on the genesis of the Bohr atom.

Even though it was a great success, the project did not unfold exactly as planned. Kuhn notes that "what [he] hadn't anticipated was the number of times people would say, 'I don't know, I can't remember that; how, why would you expect me to remember that?' In that sense, for that sort of thing, we got much less than I had hoped" (Kuhn 1997/2000, 303). Heilbron describes how they recalibrated their expectations and plan of action after they realized that the physicists' testimony was less valuable than they had initially anticipated.

> The committee [formed with representatives from the American Physical Society and the American Philosophical Society] had in mind primarily tape-recorded interviews with famous physicists. Experience showed, however, that few interviewees were reliable informants about events then as much as fifty years in the past. Kuhn and his student collaborators ... consequently devoted more and more of their effort to finding and microfilming correspondence and other unpublished manuscripts uncovered by their research. (see Heilbron 1998, 510)

Despite the challenges, a vast quantity of material was collected. In fact, as Heilbron notes: "at the close of its operations in 1964 the project had conducted and transcribed some 200 interviews with 100 informants, and had arranged for the microfilming of about 100,000 frames of material" (Heilbron 1968, 98). Heilbron provides a useful account of the contents, which gives a sense of the scale of the project, noting, for example, that "the single most important element is Bohr's Scientific Correspondence and Manuscripts, covering the years 1909–22, and occupying 18 reels of microfilm" (Heilbron 1968, 99). He also provides a useful qualitative assessment of "the letters and manuscripts in the Archive."

> They include exchanges on the same questions between several pairs of individuals. They often permit the historian to follow the development of ideas, techniques and research projects from week to week, sometimes even from day to day. They reveal ambitions, frustration, tension, rivalry,

friendship, conflict and passionate curiosity – the motive powers, as it were, of the scientific enterprise. (Heilbron 1968, 100)

In fact, the *Sources for History of Quantum Physics* is a highly valued resource. Martin Klein describes it as follows: "this collection of micro-filmed manuscripts and oral history interviews with many physicists who had taken part in the development [in early twentieth-century physics] has already become essential to anyone seriously interested in the field" (Klein 1979, 430; see also Heilbron 1968, 100).

The research conducted during this time was the catalyst for a project that would result in Kuhn's third and final monograph published during his lifetime, *Black-Body Theory and the Quantum Discontinuity, 1894–1912* (see Kuhn 1978/1987, xi).[3] The core thesis of the book is that, contrary to the popular narrative, Max Planck did not initiate the quantum revolution in 1900. Rather, Planck employed a technique, adapted from Boltzmann's work on gasses, in an effort to solve an unsolved research problem, the black-body problem. The technique, which involved assuming that radiation is emitted in discrete units, was merely a heuristic to make the problem mathematically tractable (see, Kuhn 1978/1987, viii). Kuhn notes that "even in the middle of 1906, neither restrictions on classically permissible energy nor discontinuities in the processes of emission and absorption were to be found in Planck's work" (Kuhn 1978/1987, viii). That is, Planck did not then really believe that radiation could only be emitted in discrete units.

So, to use the language of *Structure*, Planck was working in a normal scientific mode, assuming that his research results were consistent with the accepted theories. That is, he was working in classical physics. According to Kuhn, it was only later, as other physicists responded to Planck's work, that "the concept of a discontinuous physics" emerged (see Kuhn 1978/1987, ix–x). Kuhn attributes these developments to work by Albert Einstein and Paul Ehrenfest in 1906, and H. A. Lorentz in 1908. That is, they were the ones who initiated the revolution in physics. They were the ones who self-consciously made a break from classical mechanics.

Kuhn's return to history of science and historical scholarship after the publication of *Structure* is not particularly surprising, given his falling out

---

[3] Kuhn notes that this particular project, that is, understanding "when physicists first began to look for quantum conditions," began in 1972, initially, as background to a different project related to the history of quantum mechanics (see Kuhn 1978/1987, vii; see also Kuhn 1972). Kuhn also notes that "virtually all the manuscript materials referred to [in the book] were located in the course of [the History of Quantum Physics Project]" (Kuhn 1978/1987, xi).

with the philosophers at Berkeley, and his new position at Princeton (see Schuster 2018, 394). Indeed, though Kuhn was part of an HPS program at Princeton, students in the program graduated from one or other of the two participating departments, either history or philosophy (see Kuhn 1976/ 1977a, 4). As John Schuster notes, "having moved to Princeton two years after the ... publication of SSR, [Kuhn] was for the next few years mainly involved in teaching apprentice historians of science who were officially located in the great Princeton Department of History" (Schuster 2018, 394).

In fact, Kuhn's time at Princeton is not the only time he focused primarily on the history of science and set his philosophical interests aside. Kuhn notes that after he gave the Lowell Lectures, in 1951, which he described as his first attempt to write *Structure*, he realized that he "did not yet know either enough history or enough about [his] ideas to proceed toward publication. For a period ... that lasted seven years, [he] set [his] more philosophical interests aside and worked straightforwardly at history" (Kuhn 1977, xvi).

The Significance of the Black-Body Problem

Black-Body Theory and the Quantum Discontinuity was published in 1978. The reaction to the book was quite interesting and, for Kuhn, rather unexpected. Many readers were disappointed and confused that Kuhn did not bring to bear the theoretical terms for which he was famous, most importantly "paradigm" and "paradigm change," the terms that had made *Structure* so influential. Some critics felt that, as a consequence, Kuhn's analysis in the book was rather un-Kuhnian. This was captured in the title of a symposium published in the journal *Isis*: "Paradigm Lost." The symposium provided an assessment of the book from three disciplinary perspectives: history of science, philosophy of science and sociology of science.

The historian of science Martin Klein noted "the reader who wants to interpret the origins of quantum physics using the language of normal and extraordinary science, or paradigms and crises, will have to provide his own translation. Those terms, so closely identified with Kuhn's writings on the philosophy of science, are conspicuously absent from this book" (Klein 1979, 430; see also Galison 1981, 72). Klein was not at all put off by this. Rather, Klein's concerns were about the inferences Kuhn drew from the evidence. That is, he was inclined to a different interpretation of the historical facts. Most importantly, contrary to Kuhn's interpretation of the facts, Klein argues that "[he] still [does] not think that Planck's theory was 'fully classical'" (Klein 1979, 431). More precisely, Klein argues that Planck had already begun to break away from classical physics even as early as 1900.[4]

The philosopher of science Abner Shimony praised Kuhn for setting aside the concepts he had developed in *Structure* and concentrating on the primary sources relevant to the historical episode he was studying. Shimony explains that

> Kuhn deserves credit for analyzing the discovery of quantum discontinuity without any apparent control by the conceptions of scientific change expressed in [*Structure*]. The interpretations which he offers in his new history . . . seem to be the product of immersion in the scientific texts and of reflection upon scientific problems. (Shimony 1979, 437)

Thus, Shimony appreciated the fact that Kuhn kept his philosophical framework out of his historical work.

The sociologist of science Trevor Pinch agreed with Shimony's characterization of the book, but argued that "what is surprising [about the book] is that [Kuhn] has . . . provided us with a largely internal history of how discontinuity emerged in physics at the turn of the century" (Pinch 1979, 437). Pinch argues that the episode that Kuhn studies cries out for an analysis in Kuhnian terms, that is, the terms developed in *Structure*. Pinch, for example, wonders why Kuhn's insights about the complex nature of scientific discoveries do not shape his analysis in this new book (see Pinch 1979, 438). Pinch also suggests that Kuhn's analysis in *Black-Body Theory* would have been enriched if he had employed the concept of an anomaly (see Pinch 1979, 438). So, unlike Shimony, Pinch felt that this historical episode could have been illuminated by an analysis in Kuhnian terms, that is, the terms of *Structure*.

---

[4] Peter Galison provides an assessment of the debate between Kuhn and Klein regarding Planck's contribution to the quantum revolution in physics. Though Galison takes issue with aspects of Kuhn's interpretation, he does note that "even Klein . . . seems implicitly to acknowledge the strength of Kuhn's argument that Planck was not speaking of quanta in the *Lectures* . . . thus there seems to be no real dispute about Planck's lack of commitment to quantisation in 1906" (Galison 1981, 81). But Galison does not wholly agree with Kuhn either. As far as Galison is concerned, contrary to what both Kuhn and Klein assume, "in 1900–1, the question of the continuum *vs.* discreteness as such . . . was entirely peripheral to Planck's other concerns" (Galison 1981, 82). More recently, the historian of science Richard Staley has argued that "Kuhn offered convincing evidence for [his] argument . . . and Kuhn's account has been considerably refined by [other historians of physics]" (Staley 2013, 160). Indeed, Staley claims that "in the attention he paid to the relations between interpretation, mathematical techniques, and intellectual resources more generally, Kuhn has established a precisely situated intellectual sociology of the physics community of unparalleled explanatory power" (Staley 2013, 161).

Pinch also suggests "there is . . . a discernible shift in emphasis in Kuhn's current work. Science is portrayed as a process much less susceptible to human or even social influence: *nature* is firmly in the driver's seat" (Pinch 1979, 439; emphasis added). Recall from Chapter 6 that the issue of the role nature plays in resolving disputes in science was a crucial point of disagreement between Kuhn and the Strong Programme sociologists of science. Kuhn had felt that they left no role for nature. Here, Pinch is not suggesting that nature plays no role. Rather, his concern is that Kuhn seems to suggest that nature is decisive, and human and other social influences have no impact.

As a result of the apparent shift in Kuhn's focus away from so-called external factors, Pinch claims that "Kuhn . . . presents an incomplete picture of [quantum] theory's genesis" (Pinch 1979, 440). Indeed, Pinch takes this book "as the final stage of a process of retraction initiated by Kuhn in response to some of the reactions which *Structure* produced" (Pinch 1979, 440).

Like many, Pinch was struck by the fact that, faced with criticism, Kuhn seemed to qualify or soften his earlier remarks to the point that what was left was neither interesting nor contentious. Pinch, though, believed that *Structure* had been both of those things. That was what had made the book so valuable, and such a sensation! Pinch argues that Kuhn's defensive behavior was motivated by a desire to placate philosophers and historians of science. But he also argues that Kuhn's responses to criticism essentially amounted to a renunciation of all the sociological insights about science contained in or at least implied by *Structure*.

Pinch's concerns seem to anticipate the fate of Kuhn's work among historians of science, generally. History of science, it seems, took a decidedly externalist turn around the 1970s (see Kuhn 1976/1977a, 12; also 1997/2000, 305 and 319).[5] According to Kuhn, even his two closest students, Heilbron and Forman, turned away from the sort of history of science he had taught them to write (see Kuhn 1997/2000, 304). With the exception of Jed Buchwald, Kuhn notes that "all of my other students have . . . mostly . . . turned to things that are much more science and society oriented, social environment of science, institutions, and so forth" (Kuhn 1997/2000, 319). As Heilbron explains, "the approach now fashionable . . . [in the history of science] aims to integrate the social and intellectual

---

[5] While acknowledging the growth of externalist studies in the history of science in the mid-1970s, Kuhn remarks that "the philosophical import of [these studies] seems to me far more problematic than [internalist studies]," which he characterizes as being "concerned with the evolution of scientific ideas, methods, and techniques" (see Kuhn 1976/1977a, 12).

aspects of the practice of science" (Heilbron 1998, 514). Kuhn, though, did not see this change in the field as an integration. Rather, he felt that a concern for social aspects was displacing the traditional focus on intellectual aspects in the history of science.

Unlike his students, Kuhn continued to go in the same direction he had been going for years, despite the fact that the profession had changed directions significantly. Indeed, as early as 1971, Kuhn seemed to anticipate this outcome, as externalist histories of science were becoming more popular. Kuhn noted that,

> though [he welcomes] the turn to the external history of science as redressing a balance which has long been seriously askew, its new popularity may not be an unmixed blessing. One reason it now flourishes is undoubtedly the increasingly virulent antiscientific climate of these times. (Kuhn 1971/1977, 160–161)

In fact, there is some empirical support for Kuhn's worry about what was then driving historians of science to external histories of science. In a survey of graduate students in history of science programs in North America conducted in 1971, French and Gross report the following findings. First, they found that there was, among respondents, "explicit and marked concern with the social history of science and with historical interactions between science and its social context" (French and Gross 1973, 164). Second, they found 43 percent of the respondents "mentioned disenchantment with science as a motive for their involvement in history of science" (French and Gross 1973, 164). Interestingly, though, they also found that Kuhn was the most frequently mentioned author in response to the following question: "Are there any works which are models, methodologically, of the kind of scholarship you would like to do?" (French and Gross 1973, 166)

Kuhn was not only worried about the motivations behind these developments in the history of science; he also worried that if external histories came to dominate, there was a risk that "the internalities that shape the development of any [scientific] discipline" would be ignored (see Kuhn 1971/1977, 161). Indeed, Kuhn feared that his legacy in the history of science was being destroyed. No one would be left to carry on the sort of historical research he thought was important.

In 2000 Stephen Brush asked the question: "Why did Kuhn's publications in his own primary field, history of science, have so little impact on that field?", given that Kuhn "exerted a strong force on intellectual discourse in the last third of the 20th century, by the publication of a book only 200 pages long" (Brush 2000, 39). Using the *Science Citation Index*, the *Social Science Citation Index* and the *Arts & Humanities Index*, Brush gathered data on citations to Kuhn's work in general, and citations to his work in *history of science* journals specifically, for 1975, 1985 and 1995.

What Brush found was quite surprising. *Structure* was cited hundreds of times per year, ranging from 255 in 1975 to 600 in 1995. But historians of science only cited the book between seven and eleven times a year. Kuhn's historical articles were even more neglected. For example, Kuhn's 1952 paper on Boyle, his 1955 paper on Carnot and his 1961 paper on Carnot and the Cagnard engine were never cited in these three years (see Brush 2000, 45, table I). Even Kuhn's book *The Copernican Revolution* was cited in history of science journals only twice in 1975, and once in both 1985 and 1995. And the black-body theory book was cited in history of science journals only twice in 1985 and once in 1995. On the basis of Brush's quantitative study, it seems that historians really did not value or make use of Kuhn's historical scholarship. As Nathan Reingold expresses the point, "*although derived from the history of science and purporting to present a theory of how the sciences developed in time, history of science itself has remained largely immune to the Kuhnian paradigm*" (Reingold 1980, 475; emphasis in original).

Mary Jo Nye offers a different perspective on Kuhn's impact on the history of science in a reflective essay on trends in the history of physics. She claims that "Kuhn shifted many historians' interests away from a heroic history of ideas and individuals towards the history of groups and communities, that is, from histories of eureka moments of discovery to histories of communication, organization and social context" (Nye 2019, 11). Given Nye's characterization, it is no wonder Kuhn had such an impact on the sociology of science.

Incidentally, despite the response by historians to *Black-Body Theory*, Kuhn felt that it was a successful piece of historical scholarship. In fact, he claims that it was "the best study in conceptual change [he'd] done" (Kuhn 1990/2016, 28).

Brush also examined the impact Kuhn's Planck book had on physics textbooks. In particular, Brush sought to determine how many physics textbooks reflected Kuhn's view that Planck had initially regarded the quantum of energy as a mathematical assumption. Brush thought that perhaps Kuhn's work had had an impact in physics, even if it was largely neglected by historians of science. But he found that only six of the twenty-eight textbooks in physics that he examined reflected any awareness of Kuhn's view (see Brush 2000, 53, table II).

Kuhn himself suggests that the Planck book may have strained his otherwise good relations with scientists (see Kuhn 1997/2000, 282). Some scientists saw it as an attempt to give all the credit for the developments in twentieth-century physics to Einstein.

The work that Kuhn had done on Planck for *Black-Body Theory and the Quantum Discontinuity* continued to have a strong impact on his thinking, even when he returned to his research in the philosophy of science. In fact, the example discussed in the book came to play a significant role in his later discussions of scientific revolutions (see, for example, Kuhn 1987/2000). It had all the marks of the sort of episode in the history of science that he regarded as a scientific revolution. First, the change required was holistic and thus could not be made in a piecemeal way, unlike normal or cumulative changes (see Kuhn 1987/2000, 28–29). Second, there was a significant change in "the way words and phrases attach to nature" (Kuhn 1987/2000, 29), including changes "in several of the taxonomic categories prerequisite to scientific descriptions and generalizations" (Kuhn 1987/2000, 30). And, third, it "involved a central change of model, metaphor or analogy – a change in one's sense of what is similar to what, and of what is different" (Kuhn 1987/2000, 30). Certainly, by the mid-1980s Kuhn felt he had a firmer grasp of what scientific revolutions were. And this example from early twentieth-century physics was a quintessential example.

Indeed, this example also features importantly in his unpublished manuscript, *Plurality of Worlds*, the book that had been intended to be the sequel to *Structure*. Perhaps what Kuhn liked about the Planck example is that it is so far removed from the wider culture. The Copernican Revolution in astronomy, in contrast, had an impact far beyond the community of expert astronomers. It precipitated a significant change in the way Europeans saw themselves and their relationship to the cosmos. And the theoretical changes in astronomy would be entangled in Galileo's fight with the Catholic Church. In contrast, Planck's research was really of no interest to anyone outside of the community of physicists. In this way, Kuhn could keep the focus on the science. That is, he could trace the development of the scientific ideas.

Despite Brush's study of the impact of Kuhn's work in the history of science, I think one should not be left with the impression that Kuhn's historical research had little impact. It is worth considering how frequently his works in the history of science are cited. According to Google Scholar, *Black-Body Theory* has been cited over 950 times, and his article with Heilbron on the Bohr atom more than 200 times. Kuhn's *Copernican*

Revolution has been cited over 2,800 times. And this was really, principally, a textbook, rather than an original scholarly monograph. Even Kuhn's 1952 paper on Boyle has been cited more than 100 times.[6] Relatively few articles in the history of science, or any discipline, for that matter, are cited 100 times. If the numbers look unimpressive, it is only against the background that Kuhn's *Structure* has been cited more than 120,000 times.

Leaving History Behind

Not long after *Black-Body Theory and the Quantum Discontinuity* was published, Kuhn moved to MIT. Here his appointment was in the Department of Linguistics and Philosophy. His move to MIT was facilitated in part by Sylvain Bromberger, who Kuhn had known since his time at Harvard, and who was then the head of the philosophy section of the department at MIT.[7] Kuhn had nothing to do with the program in Science, Technology and Society at MIT, where some people were working in the history of science.

This move marks the end of Kuhn's active engagement with the history of science. In 1995, when he was asked about the developments in the history of science, Kuhn explained: "I'm not close enough to history of science. I mean I really, as I've gone on now in the last ten, fifteen years, really trying now to develop this philosophical position, I have just stopped reading history of science" (Kuhn 1997/2000, 322). He was mostly occupied with finishing his book, a book he never did manage to complete.

Perhaps it is worth remembering that Kuhn was ambivalent about the field of history from the start. In an interview conducted near the end of his life, he reminds people that he was not formally trained as a historian, noting that he had taken only *one* history course as an undergraduate at Harvard. He explains: "history was not a thing I thought of myself as being very fond of" (Kuhn 1997/2000, 268). Kuhn also notes that when he was an

---

[6] My searches on Google Scholar were conducted on October 8, 2020.
[7] When I was a Visiting Scholar at MIT in the fall of 2015, I was able to discuss Kuhn's hiring at MIT with Bromberger. Bromberger marveled at the fact that, down one end of the department was the office of Kuhn, the man who wrote about scientific revolutions, and down the other end was the office of Noam Chomsky, the man who caused a revolution in linguistics, and yet they never talked with each other. This was in one wing of the infamous Building 20. The building was originally built as a temporary space for radar research during the Second World War, but after the war it persisted as part of MIT until it was torn down in 1998 (see RLE Undercurrents 1997). That Chomsky caused a Kuhnian-style revolution in linguistics was noted as early as 1965, in a book review by James Thorne (see Thorne 1965).

undergraduate he "had sat in on some [history of science] lectures of Sarton's . . . and found them turgid and dull" (Kuhn 1997/2000, 275).

When Kuhn returned to America from his service during the war, he enrolled in the Ph.D. program in physics at Harvard. But right from the beginning of the Ph.D. program, Kuhn was ambivalent about a career in physics (see Kuhn 1997/2000, 273).[8] So he asked for permission to take some courses outside of physics. But it was not courses in history that he wanted to take, but rather philosophy courses (see Kuhn 1997/2000, 269 and 273). Further, Kuhn remarks: "I was not in my bones a historian; and I *was* interested in philosophy" (Kuhn 1997/2000, 275; emphasis in original).[9]

It was only when he began to work with Conant on the General Education History of Science course that he hit upon the idea of retraining as a historian of science. As he was finishing his Ph.D. in physics, he thought that a fellowship at the Harvard Society of Fellows would provide him with the means to retrain in the history of science. And this, he thought, would in turn serve his long-term goals, which he regarded as philosophical. As he explains, "I wanted to teach myself enough history of science to establish myself there in order to do the philosophy" (Kuhn 1997/2000, 276).

Kuhn's relationship with philosophy, however, was also somewhat strained. Again, a principal barrier was his lack of formal training. He had quite liked the one course in philosophy that he had taken as an undergraduate, a survey course from the Greeks to Kant (see Kuhn 1997/2000, 262–264). And, as noted above, he had decided to take some additional courses when he returned to Harvard, as he began the Ph.D. program in physics. But this attempt to retrain as a philosopher very quickly proved to be disappointing. He was not prepared to learn all that he was expected to learn if he were to pursue advanced studies in philosophy. As Kuhn reports, "yes, I was interested in philosophy, but my God, I was a graduate, I had been through war in some sense or other, I couldn't go back and sit still for that undergraduate chicken-shit" (Kuhn 1997/2000, 273). As he remarked, he knew he "wasn't going to go back and try to be a philosopher, learn to do philosophy" (Kuhn 1997/2000, 276).

---

[8] Incidentally, even if Kuhn's enthusiasm for physics was beginning to wane, as late as 1953 Van Vleck, his Ph.D. supervisor, still believed that Kuhn had a lot of promise as a physicist. In a letter from 1953, Van Vleck writes that he "would have stood up very well in theoretical research in quantum mechanics, had he wished to make this his main pursuit" (Van Vleck 1953, cited in Kaiser 2016, 77).

[9] Kuhn remarks that "there are those who feel, and feel with some justice, that I never really did get to be a historian" (Kuhn 1997/2000, 276).

Indeed, despite the profound impact *Structure* had on the philosophy of science, he was often quite detached from the profession. In fact, Kuhn notes that when he was elected the President of the Philosophy of Science Association he had *never* been a member of the association before (see Kuhn 1997/2000, 311).[10]

In many ways, Kuhn was never at home in any discipline, despite the fact that *Structure* had such a profound impact on many fields.

Given Kuhn's strained relations with both the history of science and the philosophy of science, it is worth briefly considering Kuhn's relationship with the integrated field of HPS, that is, the History *and* Philosophy of Science. While still at Princeton, Kuhn remarked that "those of you aware of my involvement with Princeton University's Program in *History and Philosophy of Science* may find odd my insistence that there is no such field" (Kuhn 1976/1977a, 4; emphasis added). The aims and methods of the two fields, he claimed, are fundamentally different (see Kuhn 1976/1977a, 4–5). Historians aim to develop a narrative, for history is "an explanatory enterprise" (see Kuhn 1976/1977a, 5). Philosophy, "on the other hand, aims principally at explicit generalizations" (Kuhn 1976/1977a, 5). As a consequence of these differences, Kuhn did not think one could work as a historian and a philosopher at the same time (see Kuhn 1976/1977a, 5).

These days, numerous scholars identify as HPS scholars. That is, they identify as working in a field that integrates both the history and the philosophy of science. And, in fact, Peter Galison claims that "Kuhn's work is . . . best seen as . . . [a] sort of continuous interaction between history and philosophy of science" (Galison 1981, 84). So, like many others, Galison considers Kuhn the quintessential scholar working in HPS.

Despite these contemporary scholars' claims that they are engaged in an integrated history and philosophy of science, there is reason to believe that Kuhn is correct in his assessment of the situation. In 2010 I conducted a study of which journals were cited in handbooks or companions in the philosophy of science, the sort of books that aim to provide an overview of the main themes and issues in the field. One would expect, if there were in fact an integrated field, that scholars contributing to such volumes would cite the leading history of science journals in their contributions. But, in fact, in the three sources I looked at there was not a single citation to the journal *Isis*, which is widely regarded as the most important journal in the

[10] Stanley Cavell recalls an incident that illustrates Kuhn's strained relationship with professional philosophy. After Cavell and Kuhn had left "an extended discussion among professional philosophers," Kuhn remarked to him: "'I wouldn't have believed it. You people don't behave like academics in any other field. You treat each other as if you are all mad'" (quoted in Cavell 2010, 354).

history of science (see Wray 2010, 428). Rather, the citations in these sources were principally to philosophy of science journals. In a follow-up study, Scott Weingart presents a co-citation network of scholars publishing in the history and philosophy of science. The diagram shows two distinct communities, separated by a sparsely populated gap (see Weingart 2015, 209, figure 2). Citing across the gap is quite uncommon. So, as much as people believe there is such a field, the evidence supports Kuhn's view. No wonder he spent his career moving between history of science and philosophy of science. There was no common meeting ground, at least not one that left strong traces in the scholarly record.

CHAPTER 8

Structure, Historicism and the History of Science

Recently, as part of the reflections on the fiftieth anniversary of the publication of *The Structure of Scientific Revolutions*, Lorraine Daston has argued that the notion of "structure" to which Kuhn appeals is "dusty and dated" (Daston 2016, 116). Though the concept was quite popular in both the social sciences and history in the early 1960s, these days the term "structure" is widely regarded as irrelevant to historical analysis. Daston argues that "most historians of science no longer believe that *any* kind of structure could possibly do justice to their subject matter. The very idea of looking for overarching regularities in the history of science seems bizarre" (Daston 2016, 117; emphasis in original). Instead, historians of science tend to focus on the particular, in recognition that the sorts of events they study are unique. Underlying this criticism is an assumption that *Structure* is or was intended to be a contribution to the history of science.

My aim in this chapter is to defend Kuhn's appeal to the notion of structure in his account of scientific change. First, I argue that *Structure* is not primarily a contribution to the history of science, despite the fact that it cites many historical sources and articles in that field. Consequently, it is a mistake to regard Kuhn's appeal to the notion of structure as a case of badly written history of science. Second, I argue that insofar as the book is a contribution to philosophy of science, the sort of thing Kuhn means by "structure" is perfectly respectable, and is often presupposed in many philosophical studies, especially in general philosophy of science. Third, I defend Kuhn's analysis of science, especially the enterprise of identifying the structure of scientific revolutions. Finally, I defend Kuhn against the charge of historicism, for the search for structures in history is often associated with a type of historicism that is widely criticized.

Is *Structure* a Contribution to the *History* of Science?

The Structure of Scientific Revolutions can be challenging to classify, just as Kuhn, the author of the book, can be challenging to classify. As mentioned earlier, Kuhn had his formal training in physics, from bachelor's to Ph.D. level. But he worked in a variety of different departments and programs, including the General Education program at Harvard, and then concurrently in the History Department and the Philosophy Department at U. C. Berkeley. After the publication of *Structure*, Kuhn worked in an interdisciplinary program in the History and Philosophy of Science at Princeton, and then in a Department of Linguistics and Philosophy at the Massachusetts Institute of Technology. Arguably, Kuhn seldom felt fully at home in any department or institutional arrangement.

During the time Kuhn was at Harvard, there were some significant radical experiments going on in the organization of the various disciplines. The social sciences, in particular, were exploring novel institutional structures. Joel Isaac has discussed this period in Harvard's history extensively (see Isaac 2012). As Isaac notes, there were a variety of institutional arrangements and structures at Harvard designed to permit and facilitate interdisciplinary work (see Isaac 2012, 6). The Harvard Society of Fellows was one part of this "interstitial academy" that cut across disciplinary lines. As a junior fellow in the Society of Fellows, Kuhn was a product of these unusual institutional arrangements.

On the one hand, it is understandable that some people have regarded *Structure* as a contribution to the history of science. The book does draw heavily on then-current historical scholarship. In fact, over three quarters of the sources cited in *Structure* are either works in the history of science, scientific biographies or scientific classics, like Darwin's *Origin* (see Wray 2015c). What is more, the manuscript was initially commissioned to be part of the *Encyclopedia of Unified Science*, as a volume devoted to the topic of the history of science (see Kuhn 1997/2000, 292; 300), and historians of science did in fact read the book with enthusiasm when it was initially published (see Kuhn 2000, 286). Charles Gillispie, a distinguished historian of science, reviewed the book in the journal *Science* (see Gillispie 1962). But, as we saw in Chapter 5, many others read the book as well, including economists, psychologists, sociologists, anthropologists, educational theorists, philosophers and scientists (see Kaiser 2016, 84, table 4.1; Wray 2017).

Further, when Kuhn wrote *Structure*, he was principally identified as a historian of science. A number of his early publications were unquestionably contributions to that field and were addressed to quite specific issues in it, like "Newton's '31st Query' and the Degradation of Gold" and "The Caloric Theory of Adiabatic Compression."[1] And his historical work is clearly situated in that genre. In fact, his major book-length contribution to the history of science, *Black-Body Theory and the Quantum Discontinuity, 1894–1912* is clearly written with historians' aims in mind (see Kuhn 1978/1987). Kuhn's intention was to understand the role and impact that Planck's research on the black-body problem had had on the revolution in early twentieth-century physics that led to the development of quantum mechanics. Indeed, as we saw in the previous chapter, the book was so historical that it disappointed his readers in the sociology of science who were hoping for a critical analysis of this episode in the history of science using the terms and concepts employed in *Structure*, that is, in terms of paradigms, anomalies and so on (see, for example, Pinch 1979; also Reingold 1980, 476). Instead, what they found was old-fashioned *internalist* history of ideas. So, it is not surprising that many thought of *Structure* as a book in the history of science.

On the other hand, though, I think Kuhn makes clear that his aims in *Structure* are not those of a *historian*. Granted, he does claim that if we study the history of science it will transform our vision of science (see Kuhn 1962/2012, 1).[2] But that is not a typical aim of historical scholarship, and Kuhn was quite aware of this. In his own reflective methodological writings on the differences between history of science and philosophy of science, he makes clear that the two fields have very different objectives or aims (see, for example, Kuhn 1976/1977a, 4–5). Whereas historians aim to develop a narrative, philosophers aim for a general theory. Further, Kuhn was insistent that you cannot do both at once. One either worked as a historian of science or one worked as a philosopher of science (Kuhn 1976/1977a, 4–5).[3] Insofar as Kuhn accurately describes the aims of the two

---

[1] Kuhn's paper "Energy Conservation as an Example of Simultaneous Discovery" (see Kuhn 1959/1977) could also be regarded as a contribution to either the history of science or the sociology of science. Merton had made multiple discoveries a key topic in the sociology of science, first in Merton (1957/1973), and then more fully in Merton (1961/1973) and Merton (1963/1973).

[2] The precise quotation is as follows: "history, if viewed as a repository for more than anecdote or chronology, could produce a decisive transformation in the image of science by which we are now possessed" (Kuhn 1962/2012, 1).

[3] Peter Galison questions whether Kuhn was successful at keeping these two enterprises separate. In fact, Galison claims that "Kuhn's work is perhaps best seen as [involving a] ... continuous interaction between the history and philosophy of science" (Galison 1981, 84). Galison suggests

fields, philosophers are much more like sociologists, at least as the latter field was widely understood and practiced in the 1960s and 1970s.

Whatever Kuhn's intentions were, contemporary historians of science generally have a bleak view of the relevance and impact of *Structure* on the history of science. As noted above, Daston makes it clear that the notion of "structure" that figures so centrally in Kuhn's analysis is outdated and has no place in the history of science. In fact, this is not a wholly new complaint against *Structure*. A similar concern was raised by Nathan Reingold in 1980. He notes that "cyclical theories of history," like the sort that Kuhn seems to offer in *Structure*, "are inherently improbable" (see Reingold 1980, 479). Elaborating he claims that "they . . . run afoul of the normal tendency of historical practice to stress particulars in time and place" (Reingold 1980, 479).

Daston further suggests that, although the book provided a blueprint of sorts for the professionalization of the history of science, *Structure* fails to measure up to the contemporary standards of historical scholarship. Kuhn thus seemed to be announcing the coming of a more professional history of science in a book that would not measure up to the new standards to which it drew attention. Peter Galison's assessment is equally damning. He describes the book as "a valiant and productive analysis of the *physics* of the 1930s done in the 1940s about the science of the seventeenth, eighteenth, and nineteenth centuries" (Galison 2016, 66; emphasis added). This is not quite an accurate description of *Structure*, for as we saw earlier, Kuhn draws extensively on the history of chemistry.

At least one influential contemporary historian of science, David Kaiser, sees past the many citations to historical sources in *Structure*. Kaiser argues that Kuhn drew the chief lessons in his book from then-recent research in psychology, not the history of science (see Kaiser 2016). Both (i) Bruner's and Postman's anomalous-playing-cards experiment and (ii) Theodor Erismann's experiments with inverted glasses play a significant role in Kuhn's presentation of his provocative theses.[4] And, in the Preface to *Structure*, Kuhn mentions that Jean Piaget's experiments on children's developing perception and understanding had a significant impact on his thinking (see Kuhn 1962/2012, xl). Kaiser even suggests that the research in psychology that Kuhn discusses may have been the *source* of Kuhn's ideas about science, and that the various historical cases discussed were merely intended to *illustrate* Kuhn's points

---

that "the unifying theme in Kuhn's writings" is "that scientific change occurs primarily through the technical applications of paradigmatic solutions to technical problems" (Galison 1981, 84).

[4] Daston discusses the inverted glasses experiment by Theodor Erismann in some detail (see Daston 2016).

(see Kaiser 2016).[5] Thus, even though Kaiser does not criticize Kuhn's historical scholarship in *Structure*, he does imply that the book is not a book in the history of science. Historians, after all, do not typically draw on research in psychology, as Kaiser alleges Kuhn does.

Kaiser's assessment of *Structure* seems to align more closely with Kuhn's self-understanding than does either Daston's or Galison's assessment. At the end of his career, Kuhn makes clear that his interests in history of science were largely instrumental. Studying the history of science was a means to advancing his philosophical project. As Kuhn explains,

> I could read texts, get inside the heads of the people who wrote them … I loved doing that. I took real pride and satisfaction in doing it. So being a historian of *that* sort was something I was quite willing to be and got a lot of kicks out of being … But my objectives in this, throughout, were to make philosophy out of it. (Kuhn 1997/2000, 276)[6]

Kuhn not only aspired to do philosophy. He also identified as a philosopher. Recall that Kuhn was very excited when he initially secured an appointment in a philosophy department at Berkeley. In his words: "I jumped at the change, because I wanted to do philosophy" (see Kuhn 1997/2000, 294). And Kuhn says that he saw the book, *Structure*, as a contribution to the philosophy of science. Kuhn explains: "my ambitions were always philosophical. And I thought of *Structure* … as being a book for philosophers" (Kuhn 1997/2000, 276 and 307).

Indeed, late in his life, when he describes the struggles he had while writing the book, it is clear that he was explicitly engaging with philosophers of science. When he spent the year at the Center for Advanced Study in the Behavioral Sciences at Stanford, he "tried to write a chapter on normal science." In his words:

> Since I was taking a relatively classical, received view approach to what a scientific theory was … I had to attribute all sorts of agreement about this, that, and the other thing, which would have appeared in the axiomatization [of the theory] either as axioms or as definitions. And I was enough of a historian to know that that agreement did not exist among the people who were [concerned]. (Kuhn 1997/2000, 296)

---

[5] Kaiser also notes that many psychologists corresponded with Kuhn about *Structure* (see table 4.1 in Kaiser 2016, 84).

[6] Obviously, Kuhn's later reflections on what he was thinking when he wrote *Structure* could be inaccurate. In fact, there is some evidence that Kuhn's memories of the past with respect to his use of the term "paradigm" changed over time (see Wray 2011, 49). Kuhn, though, seems to be aware of this.

As he was struggling to write *Structure*, at least before he hit upon the paradigm concept, Kuhn was assuming that a scientific theory was expressible in a set of sentences, from which scientists derive predictions. The view that a theory can be expressed in a set of sentences is clearly a philosophical conception of a scientific theory. Though Kuhn notes that he was enough of a historian to know that this account of a scientific theory was mistaken, he was clearly engaging with philosophers of science, and trying to develop a philosophical account of science.

Indeed, even in the Preface to *Structure* itself, he is quite explicit about his philosophical ambitions. Kuhn notes that his involvement in teaching the history of science at Harvard with Conant profoundly altered his understanding of science. His "first exposure to . . . out-of-date scientific theory and practice radically undermined some of [his] basic conceptions about the nature of science and the reasons for its special success" (see Kuhn 1962/2012, xxxix). He notes that "those [pre]conceptions were ones [he] had previously drawn partly from scientific training itself and partly from a long-standing avocational interest in the philosophy of science" (see Kuhn 1962/2012, xxxix). Thus, it was a particular philosophy of science that he was reacting against. Further, in the Preface Kuhn describes his "shift in . . . career plans, a shift from physics to history of science and then, gradually, from relatively straightforward historical problems back to the more philosophical concerns that had initially led [him] to history" (Kuhn 1962/2012, xxxix–xl). Thus, it is clear that Kuhn regards *Structure* as a contribution to the philosophy of science, not the history of science.

So it seems that those historians who have seen *Structure* as an example of poor historical scholarship are mistaken. It is not poor historical scholarship, because it was not historical scholarship at all.

Mary Hesse's 1963 review of *Structure* in *Isis*, the leading history-of-science journal, provides an interesting assessment of the book and discusses its relationship to both history of science and philosophy of science. Hesse explicitly raises the issue of how the book will be received by both historians of science and philosophers of science. She ends her review with the following remark:

> The major question for historians of science . . . is whether history bears the interpretation here put upon it [in *Structure*] . . . My own impression is that Kuhn's thesis is amply illustrated by recent historiography of science and will find easier acceptance among historians than among philosophers. (Hesse 1963, 287)

Hesse's impressions are especially interesting, given that her own work was at the intersection of the history of science and the philosophy of science. Indeed, whenever philosophers list the names of those involved in what came to be called "the historical school in philosophy of science," Hesse is almost invariably included.

Structure and Structures in Philosophy of Science

Though *Structure* is a contribution to the philosophy of science, it was not a typical one, especially for its day. Indeed, in the early 1960s, many mainstream contributions to the philosophy of science were concerned with logical analyses, with respect to either explanations or the confirmation of hypotheses. Carl Hempel's work was typical. And *Structure* is markedly different than the sort of work Hempel was doing. Consequently, it is worth considering the extent to which an analysis of science in terms of structure is an apt contribution to the philosophy of science. I aim to show that the notion of structure does have a place in philosophical analyses of science.

Paul Hoyningen-Huene briefly discusses Kuhn's use of the term "structure" in *Structure*, rightly noting that "Kuhn nowhere explicitly addresses the term's meaning" (see Hoyningen-Huene 1989/1993, 24). In fact, the term "structure" has many connotations in philosophy. In contemporary philosophy of science, it is most often associated with a family of positions under the label "structural realism" (see, for example, Worrall 1989; and Ladyman 1998). Structural realists are concerned with the structural aspects of reality that are both (i) captured by and (ii) reflected in the structural aspects of our most successful scientific theories. Maxwell's equations, for example, are alleged to capture the structural properties of light. This, though, is not the sort of structure that concerned Kuhn. The sort of structure that concerned Kuhn has its roots in *sociology*.

The structures that have been of interest to sociologists are most often deep and elusive structures. And sociologists often purport to find structure in unexpected places. In this respect, the concept "structure" is similar to other key analytic terms in sociology, including "social construction" and "function." Consider the notion of "function." As Robert K. Merton notes, it is *latent* functions, not manifest functions, that sociologists are in the business of revealing (see Merton 1949/1996). The manifest function of a social practice is related to the reason the agents engage in the practice. Latent functions are quite different. They are often related to unintended consequences. In a functional explanation a social practice is hypothesized to persist because of the effects it has, even if achieving that effect is no part of the motives of those who perpetuate the practice. For example, some social scientists have argued that the practice of extended lactation, which is common among hunter-gatherers, persists in these communities despite the fact that the hunter-gatherers are not aware of its effect on suppressing fertility, and thus controlling population growth (see Little 1991, 96–97, and Wray 2002, for discussions of this example).

Similarly, sociologists have appealed to the notion of "social construction," intending to evoke some sense of wonder at something that one might otherwise not detect or expect (on social construction, see Hacking 1999, especially chapter 1). Ian Hacking provides a long list of things social scientists (and others) have suggested are socially constructed, including women refugees, illness, serial homicide and authorship (see Hacking 1999, 1). This, I believe, is the sort of thing that Kuhn wanted to emphasize in his title: *The Structure of Scientific Revolutions*. If there is a structure to scientific revolutions, then it may not be an obvious fact.

In invoking this notion of structure, Kuhn wants to show that scientific revolutions do not happen in some random, chaotic or unpatterned way. Rather, they take on a particular form. In fact, according to Kuhn, it is not only scientific revolutions that have a structure. The development of a scientific field as a whole has a pattern or form. Once a scientific field has emerged out of the pre-paradigm stage, once all or most of the scientists working in it accept the same theory, its development takes on a certain patterned structure. Roughly, the structure is as follows. A period of normal science, in which scientists take the fundamentals of a field for granted, leads to a period of crisis caused by persistent anomalies that resist solutions. The crisis causes a slackening of the disciplinary norms and standards, which leads to the generation of new theories. The slackening of standards is essential, as the norms and standards of a normal scientific research tradition are quite rigid, and their rigidity discourages radical innovations. In fact, their rigidity is what makes scientists so effective in periods of normal science, ensuring the steady growth of knowledge that many associate with scientific progress. This is part of the essential tension in science. Finally, a new theory proves to meet the challenges the field had faced, and it becomes the dominant theory, which leads to a new normal scientific research tradition.[7]

---

[7] That Kuhn was concerned with "general patterns of scientific development" was noted by Karl Hufbauer (2012, 450).

This description of the development of the natural sciences may strike some as audacious, just as it does contemporary historians of science, like Daston. One might question Kuhn's claim that he can apprehend such regularities across so many disciplines, and over the course of hundreds of years. But I believe the view Kuhn puts forward is less audacious than one might initially think, especially if one considers the alternative.

The alternative to an account like Kuhn's is that there is *no pattern or predictability* in the development of a scientific field. That is, each scientific field develops in a unique way, and even the changes within a single field are unlike earlier changes in that field. If this were the case, then there would be no pattern or structure to apprehend. Of course, some things may happen with no predictable regularity, or follow no set pattern. But sociologists have uncovered surprising patterns in many sorts of phenomena.

Émile Durkheim, for example, discovered robust patterns in suicide rates in a number of European countries throughout the second half of the nineteenth century (Durkheim 1930/1951). This was a tumultuous time in European history, affected by numerous radical changes due to industrialization, increasing urbanization and significant population growth. Societies were far from stable. But Durkheim did find resilient seasonal patterns in suicide rates. Suicides were more common in the summer, when the days were long. He also found resilient patterns with respect to marital status and suicide rates. Marriage reduced the suicide rates of men but increased the suicide rates of women. And there were also patterns with respect to religious affiliation and suicide rates as well. Catholics are far less prone to commit suicide than are Protestants. Studies like Durkheim's should heighten our sensitivity and openness to finding a pattern of some kind in any sort of social phenomena.

So there is some prima facie plausibility to Kuhn's claim that revolutionary changes of theory in science happen according to some sort of pattern. Scientific research, after all, is conducted by social groups, and thus prone to the same sorts of dynamics that affect other social groups. If there are patterns in suicides, there may well be patterns in the development of scientific fields.

But perhaps science is different, and there is no pattern or structure to it. Perhaps it is not like other sorts of social phenomena that have a structure. This is somewhat doubtful. Further, it is worth noting that the project of looking for a structure in ***scientific change*** neither began with nor ended with Kuhn.

Henri Poincaré, for example, suggested that as scientists make increasingly more penetrating investigations of nature, a structure or pattern emerges in the way they conceive of it. Specifically, Poincaré suggests that as scientists seek to develop a better understanding of the world they "discover the simple beneath the complex, and then the complex from the simple, and then again the simple beneath the complex, and so on, without ever being able to predict what the last term will be" (see Poincaré 1903/2001, 114). Even Karl Popper, who disagreed with Kuhn on many matters, believed that the sciences develop according to a pattern (see Popper 1975/1981). Popper compares scientific change to evolutionary change in the biological world. Just as biological species continuously face new challenges and may even be driven to extinction if they do not adjust to the changes in their environment that are the source of these challenges, scientific theories are constantly facing new tests which threaten to falsify them, if they cannot be modified in ways that enable them to meet the threat. Popper conceived of this as an iterative and endless process.

Also, more recently, Andrew Abbott has argued that the social sciences, unlike the natural sciences, have a developmental structure that resembles a fractal pattern, driven by the recycling or reinvention of older concepts and distinctions (see Abbott 2001). In sociology, for example, a period in which conflict theorists are the dominant group gives way to a period in which consensus theorists dominate. And this period is followed by a period in which conflict theorists dominate, etc., with each new iteration absorbing something from its predecessors (see Abbott 2001, 17). So the case for there being a structure to scientific change is less dubious than Daston implies. At any rate, it seems less dubious to social scientists and philosophers of science. Granted, it may not be the sort of thing that interests historians or that has a place in a proper historical analysis of science, but that such a structure is there to be found is at least prima facie plausible.

It is not just the macro-structure of scientific change that Kuhn thought had a pattern or structure. He also insisted scientific discoveries have a structure (see Kuhn 1962/2012, chapter 6). As Kuhn explains, "discoveries ... are not isolated events but extended episodes with a regularly recurrent *structure*" (see Kuhn 1962/2012, 53; emphasis added). Here he is quite explicit that he is looking for a regular recurrent structure or pattern. He describes the structure in detail:

> [1] discovery commences with the awareness of anomaly ... [2] it then continues with a more or less extended exploration of the area of anomaly. And [3] it closes only when the paradigm theory has been adjusted so that the anomalous has become the expected. (Kuhn 1962/2012, 53; numerals added)

Kuhn is clear that this is a general account of discovery. He notes, he is providing "an elucidation of the *nature* of discoveries" and "an understanding of the circumstances under which discoveries can come about" (Kuhn 1962/2012, 57; emphasis added).

Kuhn also discussed the structure of scientific discoveries in an article published in the journal *Science* just prior to the publication of *Structure*. There, Kuhn is even more explicit about there being a structure to scientific discoveries. "It is ... just because [discoveries of the sort that concern Kuhn] demand readjustments [to the accepted theory] ... that the process of discovery is necessarily and inevitably one that shows *structure* and that therefore extends in time" (Kuhn 1962, 764; emphasis added). The notion of structure is thus integral to his project. It was not an accident that he chose that term.

I think there are very good grounds for supporting Kuhn's theory of scientific change. First, as Kuhn notes, every theory is a partial representation of reality. That is, every theory directs attention to some variables and not others. This, according to Kuhn, is why scientists working in a normal scientific tradition are generally so efficient in addressing their research goals (see, for example, Kuhn 1962/2012, 36). The theory they accept, the theory through which they have been socialized to see the world, will lead them to attend to only those features of that world that are regarded as salient, where what is "salient" is determined by the theory. But insofar as a theory is partial, ultimately it is bound to be extended or applied to phenomena for which it is ill suited to model. This seems to be an inevitable result of applying a theory too broadly, that is, to phenomena it was neither designed nor fit to account for. Because of the partial nature of theories, every theory is bound to fail at some time (see Wray 2019). Consequently, the sort of cycle of change that Kuhn describes is to be expected. New theories, theories that make fundamentally different assumptions about reality than the theories we currently accept, will replace our current theories. Only then will scientists be able to effectively investigate the phenomena they seek to understand.

Kuhn's focus on the *structure* of scientific revolutions has not been completely overlooked by commentators. Both Alexander Bird (2015, 37) and Ian Hacking have picked up on this. In the Introductory Essay to the fiftieth anniversary edition of *Structure*, Hacking notes that "*structure* and *revolution* are rightly put in the book's title" (see Hacking 2012, x; emphasis added). As Hacking explains, "Kuhn thought not only that there are scientific revolutions but also that they have a structure. He had laid out the structure with great care, attaching a useful name to each node of the structure" (Hacking 2012, x). Hacking, though, also recognizes that historians do not think of history as having a structure. So, he is attuned to Daston's concern. Hacking attributes Kuhn's penchant for looking for a structure in scientific change to his early training in physics (see Hacking 2012, xi and xxxiii). As Hacking explains, "Kuhn's instinct as a physicist . . . led him to find a simple and insightful all-purpose structure" (see Hacking 2012, xi).

I am inclined to disagree with Hacking about the source of Kuhn's thinking about structures in science. As Daston recognizes, when Kuhn was writing *Structure*, the notion of structure was a popular concept in sociological analyses. Indeed, in America, in the 1950s and 1960s, structuralism (and functionalism) may have been the dominant paradigm in sociology. Merton, the father of American sociology of science, developed analyses of science that explained much in terms of the structure of scientific research communities, including, for example, priority disputes and multiple discoveries. Kuhn was very familiar with Merton's work, including Merton's dissertation, *Science, Technology, and Society in Seventeenth-Century England*, which Kuhn had read when he worked with James B. Conant on the General Education Natural Science course at Harvard (see Kuhn 2000, 279; 287; see also Kuhn 1949). Further, as we saw earlier, Kuhn was also in contact with Merton during the late 1950s and early 1960s (see Cole and Zuckerman 1975, 159). In fact, there are a number of letters back and forth between Kuhn and Merton in the Thomas S. Kuhn Archives at MIT. These include an exchange about simultaneous discoveries, the concept of an anomaly, and Kuhn's measurement paper (see, for example, Kuhn 1958; Merton 1959; and Kuhn 1959c). Merton even expressed a willingness to write a supportive letter on Kuhn's behalf to the University of Chicago Press if Kuhn encountered any difficulties in getting *Structure* published (see Merton and Barber 2004, 267; also Cole and Zuckerman 1975, 159).

Further, in the Preface to *Structure*, Kuhn explicitly notes the influence of sociology on his thinking. He explains that, while he was a fellow at the Harvard Society of Fellows, two events led him to realize that his ideas about scientific change "might [need] to be set in the sociology of the scientific community." The first was reading Ludwik Fleck's book, *Genesis and Development of a Scientific Fact*, and the second was a discussion with the sociologist Francis Sutton (see Kuhn 1962/2012, xli). Kuhn repeated similar sentiments in the foreword to the English translation of Fleck's book. He explains there that "Fleck's text helped [him] to realize that the problems which concerned [him] had a fundamentally sociological dimension" (see Kuhn 1979, viii). Thus, Kuhn's penchant for finding structures was more likely due to his familiarity with sociology of science rather than his training in physics. And, as mentioned earlier, already in 1952, Philipp Frank had asked Kuhn to be part of a "research project under the general title 'sociology of science,'" a project that Frank was pursuing under the auspices of the Institute for the Unity of Science (see Frank 1952). This was the year after Kuhn gave the Lowell Lectures, "The Quest for Physical Theory: Problems in the Methodology of Scientific Research," his first attempt to write *Structure* (see Kuhn 2000, 289; also Galison 2016, 59). Thus, a sociological orientation was on Kuhn's mind quite early in the writing of *Structure*. Indeed, he refers to the developing sociological dimension of his project in a letter to the Director of the Center for Advanced Study in the Behavioral Sciences, written in 1959, following his stay at the Center (see Kuhn 1959a).

What Structure Is *Not*

It is worth distinguishing Kuhn's view of scientific change from the view that Popper calls "historicism," for I suspect that some of Kuhn's critics may be attributing some form of historicism to Kuhn.

Popper characterizes historicism as the view that "*there are to be found in history general laws, rhythms, or patterns. And with these the social sciences can make predictions about the future*" (see Reynolds 1999, 277; emphasis in original). Popper, though, argues against "the possibility of predicting historical developments to the extent to which they may be influenced by the growth of our knowledge" (see Popper 1957/1991, vii).[8] That is, he insists that we cannot anticipate the future, when the future depends on the growth of our knowledge. Consequently, we cannot expect to make accurate predictions about the future.

Popper clarifies that he does not

> refute the possibility of *every kind* of social prediction; on the contrary, [his argument against historicism] is perfectly compatible with the possibility of testing social theories – for example, economic theories – by way of

---

[8] Andrew Reynolds notes that Popper's characterization of historicism is at odds with another popular characterization of historicism associated with Vico, Dilthey and Collingwood (see Reynolds 1999, 276–277). Given Popper's characterization of historicism, the social scientist and historian will employ the same methods as the natural scientists. In contrast, Vico and others characterize historicism as the view that "*history has its own methods which are distinct from those of the natural sciences*" (Reynolds 1999, 276; emphasis in original).

predicting that certain developments will take place under certain conditions. (Popper 1957/1991, vii; emphasis added)

Thus, as far as Popper is concerned, general trends may be predictable. What is not predictable, he insists, are specific future outcomes.

I suspect that underlying Daston's resistance to appeals to the notion of structure is a presumption that an explanation about scientific change in terms of structure, that is, an explanation of the sort that Kuhn offers, is inevitably linked to implausible historicist claims about science and scientific change. Whether in fact Daston believes this or not is irrelevant, as at least one commentator has connected Kuhn's view with the type of historicism that Popper objects to. Andrew Reynolds claims that "it would not be entirely strange to [classify] Kuhn's thesis of scientific revolutions" as a form of "Popperian historicism," that is, the sort of historicism that Popper attacks (see Reynolds 1999, 277).

But I think this is a mischaracterization of Kuhn's view. Kuhn certainly does not link his search for structures or patterns with an expectation that we can predict the future development of science. Indeed, although Kuhn thinks that there is a pattern or structure to the development of science, there is good reason to believe that he does not accept Popperian historicism. After all, Kuhn is adamant that science is not aptly described as going *anywhere* (see Kuhn 1962/2012, 169–172; see Wray 2011, chapter 6).[9] As far as Kuhn is concerned, there is no final goal, like the "true" account of reality, to which science aims, that could explain the development of science. So, for this reason it would be a mistake to think of Kuhn's view as a form of historicism, at least as Popper understood the term. Further, nowhere does Kuhn imply that he can anticipate the future developments

---

[9] Shearmur notes that Popper was "also critical ... about there being a 'plot' to history, and of historical periodization" (see Shearmur 2017, 57). Clearly, in one sense, Kuhn is offering a periodization of sorts in his account of the development of scientific disciplines or fields. In fact, Kuhn recognizes this (see Kuhn 1969/1977). Kuhn's most explicit discussion of periodization is in a paper titled "Comments on the Relations of Science and Art." There, Kuhn remarks that one of his innovations with respect to philosophy of science was to apply the notion of periodization to science. He has in mind here the pattern of development that alternates between (i) "periods during which practice conforms to a tradition based upon one or another stable constellation of values, techniques, and models," and (ii) "periods of relatively rapid change in which one tradition and one set of values and models gives way to another" (Kuhn 1969/1977, 348). Kuhn notes that it has long been recognized that this sort of developmental pattern applies to the arts and philosophy (see Kuhn 1969/1977).

There is good reason to believe that Popper did not think Kuhn was offering the sort of periodization that he objects to in his attack on historicism. After all, Popper never discusses Kuhn's view in these terms, and he had plenty of opportunity to do this, including in his contribution to the Bedford College conference organized by Imre Lakatos, when he chose to take on Kuhn's notion of normal science (see Popper 1970/1972).

in science. He is quite explicit that neither he nor anyone else knows where science is going.

Nonetheless, Kuhn insists that we can explain the growth of scientific knowledge without any appeal to teleology. Indeed, when he wrote *Structure*, Kuhn believed that this was one of the most contentious claims he made in it. What is more, he expected his remark that science is pushed from behind rather than driven toward the truth would meet with significant resistance (see Kuhn 2000, 307). But that aspect of his view was largely neglected.

Significantly, the sorts of patterns that Kuhn purports to have identified are not attached to any sort of specific conceptual developments in any of the sciences. Thus, one cannot expect to predict precise changes of theory. Kuhn makes no pretentions to being able to anticipate future conceptual developments. One cannot even expect to predict the timing of the next change of theory. What Kuhn purports to be able to predict is the development of scientific fields in some sort of general form. But, significantly, if his theory of science is correct it certainly tells us a lot about science and scientific knowledge. It highlights a pattern of change in science, and provides us with significant insight into what we can expect from the natural sciences. For example, it suggests that the theories we currently accept will be replaced in the future by theories that make significantly different assumptions about reality. But what exactly these new assumptions will be, no one can know in advance.

Incidentally, Friedrich Hayek makes a similar claim about the predictive power of theories in the social and biological sciences (see Hayek 1964). He characterizes them as sciences about complex phenomena. And Hayek certainly thought that even this degree of predictive power was useful and insightful.

Recently, Alexander Bird (2015) has argued that there are two historicist strains in Kuhn's historiography of science. First, Bird claims, Kuhn is committed to "a *conservative* strand of historicism" according to which "the evaluation of a theory is relative to a specific tradition of puzzle-solving" (see Bird 2015, 25 and 26). This was quite contrary to the long-standing tradition in philosophy of science that sought to identify timeless and universal criteria for theory evaluation. Second, Bird claims, Kuhn is committed to a "*determinist* strand" of historicism, according to which "there is a ... fundamentally cyclical pattern [to the development of science] with ... alternating phases of normal and extraordinary (revolutionary) ... science" (see Bird 2015, 25 and 27). Central to Kuhn's cycle are disruptive revolutionary changes of theory that involve incurring some loss of explanatory power relative to the long-accepted theory. Bird claims that this, too, is at odds with a long-standing tradition in philosophy of science that regards the progress of science as strictly cumulative, with no setbacks. Significantly, Bird's characterization of the two dimensions of Kuhn's historicism is not the same as the sort of historicism that Popper was criticizing.

In summary, my aim has been to vindicate Kuhn in his appeal to the notion of structure in his account of science. First, I have argued that Kuhn did not write *Structure* as a contribution to the history of science. Consequently, whether or not the notion of structure has a place in historical analyses of science is irrelevant to assessing Kuhn's book. Second, I have argued that the notion of structure has a legitimate place in the philosophy of science. As in sociological studies that appeal to structure, a philosophical investigation that appeals to the notion of structure signals a concern to get at a pattern in the phenomena. And there is no compelling reason to think that there is not a structure or pattern in the way that scientific fields change. Third, I have tried to provide some reason to believe that Kuhn's appeal to structure in his analysis of the dynamics of theory change is plausible. Given the partial nature of our theories, it is no wonder that they are ultimately rejected, and replaced by new theories that make significantly different assumptions about reality. Finally, I have argued that the sort of structural explanation Kuhn offers does not commit him to some sort of implausible historicism of the sort that Popper criticized.

PART IV

Kuhn's Philosophical Legacy

In this Part I examine Kuhn's philosophical legacy. First, I examine the initial responses by philosophers of science to *Structure*. A particular reading of *Structure* had become widely accepted among philosophers quite early on. It portrayed Kuhn as a relativist, and as denying the rationality of science. This reading of Kuhn became a convenient foil in the philosophy of science, a view against which others would present their own view. Second, I examine a lasting and significant impact Kuhn had on a topic that he did not actively or explicitly engage with. Specifically, I examine how Kuhn's *Structure* contributed to shaping the contemporary realism/anti-realism debates in philosophy of science. Though Kuhn did not engage in these debates, I argue that he set the agenda with the problem of theory change, a consequence of reflecting on the nature of scientific revolutions.

CHAPTER 9

Squeals of Outrage from Philosophers

As we saw in the previous chapter, Kuhn conceived of *Structure* as a contribution to the philosophy of science. But it seems that Kuhn anticipated that the book might be ignored or misunderstood by philosophers of science. In a letter to James B. Conant from 1961, just one year before the publication of *Structure*, Kuhn remarked that

> I share your conviction that the book, if at all successful, will raise some dust. If there are no squeals of outrage, I shall have failed in part of my objective. In particular . . . I expect the philosophers, more than any other concerned professional group will brush me aside. They have other holes in which to put my pegs, and they will conclude from the fact that my pegs are square only that I am no philosopher. (Kuhn 1961b)

It is quite probable that Kuhn had forgotten about these remarks by the time *Structure* was in print. But he was right. There were squeals of outrage. And many philosophers thought that Kuhn's pegs were square.

In fact, among philosophers of science, the fate of *Structure* was sealed, more or less, by 1970. The year 1970 is a convenient date to pick for two reasons. First, it was the year that *Criticism and the Growth of Science* was published. This is the collection of essays originally presented at the 1965 conference at Bedford College where Thomas Kuhn and Karl Popper faced off against each other, giving rise to what came to be called "the Kuhn–Popper debate," which Steve Fuller has described as nothing less than "the struggle for the soul of science" (see Fuller 2004; see also Worrall 2002, 65–66). Second, it was also the year that the second edition of *Structure* was published, which included the Postscript, in which Kuhn attempted to clarify his position in light of the first round of criticisms.[1]

This first round of criticisms was, to a large extent, also the final round from philosophers of science. That is, there is a certain sense in which

---

[1] As Kuhn explains, the "postscript was first prepared . . . for inclusion in [the] Japanese translation of [the] book" (Kuhn 1969/2012, 173, fn. 1).

neither Kuhn nor his philosophical critics really got past these initial criticisms. Some of these concerns proved to be problems that stuck with Kuhn throughout his life and prevented him from moving on to new projects. Some of the concerns were catalysts for ongoing debates in the philosophy of science that are still alive today. But, significantly, a reading of *Structure* ossified quite quickly, and it has remained with philosophers ever since, despite Kuhn's many attempts to defend, clarify or develop his view in later writings. Many philosophers were content to have the Kuhnian view, as initially interpreted in the 1960s, even if Kuhn himself did not endorse the view attributed to him. Kuhn's *Structure* would become a convenient foil for many philosophers of science (see, for example, Longino 1990).

In fact, one Kuhnian view did take on a life of its own. It was manifest in the Strong Programme's reading of *Structure*, which Kuhn continued to distance himself from throughout his career.[2] By the mid-1980s, Larry Laudan would attack an interpretation of Kuhn's philosophy created by what he referred to as Kuhn's expositors (see Laudan 1984, xiii). After acknowledging the profound influence Kuhn had had "on our perspective on science," Laudan notes that, "less frequently admitted is the fact that, in the twenty-two years since the appearance of *The Structure of Scientific Revolutions*, a great deal of historical scholarship and analytic spadework has moved our understanding of the processes of scientific rationality and scientific change beyond the point where Kuhn left it" (Laudan 1984, xii). Laudan was dismayed that philosophers of science continued, even in the mid-1980s, to wrestle with this Kuhnian picture of science, one that even Kuhn himself did not endorse.

In this chapter, I want to examine the nature of the views that came to be attributed to Kuhn, as well as what Kuhn said in response to the criticisms associated with this interpretation.

The Early Criticisms

It is worth dividing the early criticisms into four themes: (i) concerns about the paradigm concept; (ii) concerns about scientific rationality; (iii) concerns about relativism; and (iv) concerns about whether Kuhn's account is normative or descriptive. Let us examine each of these in turn.

---

[2] For example, in the Preface to *Essential Tension*, Kuhn suggests that the new sociologists of science, who "informally [describe] themselves as 'Kuhnians,'" are "seriously misdirected" (see Kuhn 1977, xxi). As we saw in Chapter 6, Kuhn would continue to repeat similar concerns (see, for example, Kuhn 1992/2000, 106).

Paradigms

First, there are those who raised concerns about the paradigm concept. This was the focus of Dudley Shapere's review of the book in the journal *The Philosophical Review*. Shapere argued that Kuhn's

> view is made to appear convincing only by inflating the definition of "paradigm" until that term becomes so vague and ambiguous that it cannot easily be withheld, so general that it cannot easily be applied, so mysterious that it cannot help explain, and so misleading that it is a positive hindrance to the understanding of some central aspects of science. (Shapere 1964, 393)

This is quite a criticism. Shapere tells us that the term "paradigm" is vague and ambiguous, too general, mysterious and misleading. The influence of Shapere's reading should not be underestimated. According to Google Scholar, his review has been cited 450 times.[3] And remember, this is not a scholarly journal article. It is merely a *book review*! Shapere would review the second edition of *Structure* that was published in 1970, alongside Lakatos' and Musgrave's *Criticism and the Growth of Knowledge*, in the journal *Science*, under the title "The Paradigm Concept" (see Shapere 1971).

The paradigm concept was also the focus of Margaret Masterman's paper at the Bedford College conference in 1965. Masterman had counted *all* the various ways in which Kuhn used the term "paradigm" in *Structure* (see Masterman 1970/1972, § 1). Masterman was a bit of an outsider at the conference, as many of the other contributors were Popperians or associates of Popper. In fact, she saw herself as providing an antidote to the excessive fawning over Popper, and claimed to be injecting "a little pro-Kuhnian aggressiveness ... into [the] symposium" (Masterman 1970/1972, 61). Masterman was quite critical of Popper's research program. She argued that Kuhn's "paradigm view ..., by successfully establishing the characteristic scientificness of science, successfully combats the aetherial philosophicness of the Popperian 'falsifiable metaphysics' view" (see Masterman 1970/1972, 67). In fact, Masterman had been a witness to Popper's other famous face-off, with Ludwig Wittgenstein, which ended with Wittgenstein waving a fire-poker at Popper and then storming out of the room (see Edmonds and Eidinow 2002).

Masterman was an outsider in another sense. She was not a philosopher, and she made a point of emphasizing this. She claimed to work "in the

---

[3] I conducted this Google Scholar search on July 21, 2020. This book review has been cited more than any other publication by Shapere.

computer sciences" (see Masterman 1970/1972, 60), and worked at the Cambridge Language Research Unit (see Masterman 1970/1972, 59). The fact that she was a working scientist, she explained, was why she appreciated Kuhn's book in a way that others did not. In her words, "Kuhn has really looked at actual science, in several fields, instead of confining his field of reading to that of the history and philosophy of science" (Masterman 1970/1972, 59).

Though Masterman was critical about the undisciplined way in which Kuhn used the paradigm concept, in the end she was very sympathetic to his project. The second half of the first sentence in her paper is unequivocal. She notes that her paper "is written on the assumption that *T. S. Kuhn is one of the outstanding philosophers of science of our time*" (Masterman 1970/1972, 59; emphasis added). This assessment of Kuhn, it seems, has been largely lost on those who cite Masterman's paper. Her paper is most often cited to note the problems with the paradigm concept, and the fact that she identified twenty-one different uses of the term "paradigm" in *Structure*.

Rationality

Other early philosophical critics focused on the alleged irrationality implied by Kuhn's view. Imre Lakatos, for example, argued that Kuhn's view made scientific change a matter of mob psychology (see Lakatos 1970/1972, 178). This expression in particular is repeated ad nauseam (see, for example, Matthews 2003; Gutting 2003; and Gattei 2008). According to Lakatos,

> for Kuhn scientific change – from one "paradigm" to another – is a mystical conversion which is not and cannot be governed by rules of reason and which falls totally within the realm of the (*social*) *psychology of discovery*. Scientific change is a kind of religious change. (Lakatos 1970, 93; emphasis in original)

Lakatos also claims that, according to Kuhn, a paradigm change involves "a bandwagon effect" (see Lakatos 1970, 178). Lakatos seems to be assuming that any sort of explanation of scientific behavior that appeals to psychological factors entails a denial of rationality, an assumption that many philosophers make. However, it is not the sort of assumption that sociologists of science make, especially the proponents of the Strong Programme, who, as we saw in Chapter 6, believe that all beliefs, true and false, have social causes.

Lakatos' critique of Kuhn was also a vehicle for presenting his own view, so-called sophisticated falsificationism, which he contrasts with what he calls "Popper's naïve falsificationism." So, Lakatos' paper was as much aimed at taking Popper down as it was at criticizing Kuhn.

Whereas Lakatos focused on the alleged irrationality of revolutionary science, Popper and others at the conference suggested that even Kuhn's notion of normal science was a threat to scientific rationality.[4] Popper, for example, argues that "the 'normal' scientist, as Kuhn describes him, is a person one ought to be sorry for ... The 'normal' scientist ... has been taught badly" (Popper 1970/1972, 52). Elaborating, Popper explains that the normal scientist "has been taught in a dogmatic spirit: he is the victim of indoctrination. He has learned a technique which can be applied without asking for the reason why" (Popper 1970/1972, 53). Finally, Popper notes, the normal scientist "is, as Kuhn puts it, content to solve 'puzzles'" (Popper 1970/1972, 53). Popper thus compares the normal scientist to the engineer.[5]

The sort of behavior that Kuhn ascribes to normal scientists is antithetical to Popper's ideal of a scientist. According to Popper, a scientist should be prepared to relinquish any belief. Our theories always retain their hypothetical quality. On Popper's view, the scientist's allegiance is to the facts, which, though incapable of proving theories true, can decisively show us which of our theories and beliefs are false.

Indeed, philosophers were not the only ones concerned with Kuhn's characterization of normal science. Leonard Nash, the chemist with whom

---

[4] John Watkins' paper was also directed against normal science (see Watkins 1970). Feyerabend examines normal science as well, both (i) from a normative point of view – *should* scientists pursue their research goals as Kuhn suggests they do in phases of normal science? – and (ii) from a descriptive point of view – does the sort of theoretical monism that Kuhn says characterizes normal science in fact characterize scientific practice? (See Feyerabend 1970, § 4–6).

[5] Popper's fight with Wittgenstein was over puzzles as well. As Popper explains, "early in the academic year 1946–47 I received an invitation from the Secretary of the Moral Sciences Club at Cambridge to read a paper about some 'philosophical puzzle.' It was of course clear that this was Wittgenstein's formulation, and that behind it was Wittgenstein's philosophical thesis that there are no genuine problems in philosophy" (see Popper 1974/1992, 140).

Kuhn notes that "when [he describes] the scientist as a *puzzle solver* and Sir Karl describes him as a *problem solver* ... the similarity of [their] terms disguises a fundamental divergence" (Kuhn 1970/1977, 271; emphasis added). According to Kuhn, when a scientist fails to solve a research puzzle, it is the scientist who "is in difficulty, not current theory," whereas, according to Popper, when a scientist fails to solve a problem, it is the theory that is impugned (see Kuhn 1970/1977, 271, fn. 6). Thus, Popper claims that we have learned to "let our conjectures, our theories die in our stead" (see Popper 1978, 354). Kuhn also insists that Popper underestimates the power of the puzzles that characterize normal science. He explains that "in the developed sciences ... it is technical puzzles that provide the usual occasion and often the concrete material for revolution" (Kuhn 1970/2000, 141). That is, it is in the course of normal research, guided by the accepted theories, that scientists encounter anomalies that prove resistant to normalization and thus start the process that results in a revolutionary change of theory.

Kuhn taught the history of science course at Harvard after Conant gave it up, raised a similar concern. In a letter to Kuhn, shortly after the publication of *Structure*, Nash wrote: "I feel impelled to struggle against your concept of science as a game of problem solving because . . . I want strongly to resist anything that seems to reduce the dignity of science as a way of knowing the world" (Nash 1963).[6] Thus, like Popper, Nash thought Kuhn's description of normal science diminished the dignity of science. Nash seems especially prescient here, anticipating the now widely accepted association of Kuhn with postmodernism.

If these critics are correct, then on Kuhn's view science is rational neither in its normal science phases, nor in its revolutionary phases.

Relativism

A third, related, line of attack against Kuhn was one that accused him of advocating or defending some form of relativism. This is the line of argument developed by Israel Scheffler, among others. Significantly, this line of attack is not limited to the concern that Kuhn undermines the rationality of science. Rather, it is a far more global skepticism. After explaining Kuhn's view, Scheffler notes:

> now we see how far we have come from the standard view. Independent and public controls are no more, communication has failed, the common universe of things is a delusion, reality itself is made by the scientist rather than discovered by him. In place of a community of rational men following objective procedures in the pursuit of truth, we have a set of isolated monads. (Scheffler 1967, 19)

All of this, Scheffler claims, is implied by Kuhn's philosophy of science. Scheffler, though, notes that "[he] cannot . . . believe that this bleak picture, representing an extravagant idealism, is true" (Scheffler 1967, 19).[7] Indeed, many were unwilling to accept the picture of science that Kuhn seemed to present.

This reading of Kuhn was not unique to Scheffler. David Papineau casts both Kuhn and Feyerabend as relativists in his book review of Mary Hesse's

---

[6] Nash also felt compelled "to emphasize . . . the element of continuity in scientific revolutions – and to . . . minimize the element of discontinuity that you [that is, Kuhn] understandably want to stress" (Nash 1963).

[7] Scheffler's book *Science and Subjectivity* has been cited 1,172 times. It is Scheffler's most cited publication. I conducted this Google Scholar search on July 21, 2020.

The Structure of Scientific Inference (see Papineau 1974, 167). Papineau explains that

> for Kuhn and Feyerabend theoretical commitments play a decisive role in scientific development, not only in focussing the scientist's attention on particular problems and suggesting solutions, but also by *infecting* the interpretation he gives to all his concepts, including those used to describe observational evidence. (Papineau 1974, 167; emphasis added)

The term "infect" makes it clear that the effects of a scientist's theoretical preconceptions on their interpretation of observational evidence are, without a doubt, pernicious. That is, scientists are prone to develop a distorted picture or understanding of the world.

Indeed, Kuhn tells a story of a conversation he had with his colleague Carl Hempel, after Hempel had returned from a conference. According to Hempel, people were saying "'that book [that is, *Structure*] should be burned!' and 'All this talk about irrationality! . . .' Irrationality in particular, irrationality and relativism" (Kuhn 1997/2000, 307). Clearly, there were indeed squeals of outrage following the publication of *Structure*.

The Normative and the Descriptive

A fourth popular early line of attack against *Structure* was that Kuhn was unclear with respect to whether his aims were normative or descriptive. Paul Feyerabend raises this concern in the beginning of his contribution to the Bedford College conference. As he explains:

> whenever I read Kuhn, I am troubled by the following question: are we here presented with methodological prescriptions which tell the scientist how to proceed; or are we given a description, void of any evaluative element, of those activities which are generally called "scientific"? (Feyerabend 1970, 198)

According to Feyerabend, "Kuhn's writings . . . do not lead to a straightforward answer" (Feyerabend 1970, 198).[8]

---

[8] Despite Feyerabend's criticism of Kuhn's philosophy of science, he had a great appreciation for his *historical* scholarship. In a letter to Lakatos, Feyerabend remarks: "I am rereading Kuhn's *The Copernican Revolution*. A marvellous book – I wish he were still writing in the style he was using then" (see Feyerabend 1968/1999, 142). And in another letter to Lakatos, Feyerabend notes that "Heilbron and Kuhn have written a paper on atomic models, and it came out in the newly founded *Journal of Historical Studies in the Physical Sciences* . . . in the same number Forman has an article . . . Again, very good. The Kuhnians are doing *very interesting* historical work" (Feyerabend 1969/1999, 181). The journal in which these papers are published is actually titled *Historical Studies in the Physical Sciences*. Incidentally, Kuhn was also very impressed with Forman's paper, as the following story makes clear. "I am still a devotee of Paul's *Weimar Culture and the Quantum Theory* . . . I remember

The key issue here is what to make of Kuhn's philosophy of science. Only if it has normative implications is it relevant to the *philosophy of science*, as traditionally understood. But it was unclear to Feyerabend and other readers (i) whether Kuhn intended to draw normative implications from his analysis, and (ii) whether drawing such implications was warranted. Critics could thus grant that Kuhn may be accurately describing how scientists sometimes, or even mostly, behave, but that is quite irrelevant to the normative question of how they should behave, which is the proper purview of the philosopher of science. Indeed, the narrow focus of the Logical Positivists on the logic of science was due to their belief that only logic, broadly construed to include both deductive and inductive logic, was relevant to the philosophy of science, for only logic was relevant to understanding what scientists should and should not believe.

An investigation of the psychological or sociological factors that influence scientists may be interesting. But many questioned the relevance of such factors to a philosophical understanding of science. Indeed, the distinction between the context of discovery and the context of justification was intended to keep the boundary clear between properly philosophical considerations, like justification, confirmation and warrant, and philosophically irrelevant considerations.

The Fallout

I now want to discuss some of the fallout and lasting effects of each of these lines of criticism of Kuhn's view. This will enable us to assess the impact of *Structure* on subsequent philosophy of science. I will also draw attention to some of the directions in which Kuhn developed his philosophy, even if these developments were often ignored by his readers.

Kuhn would spend the first ten or so years after the publication of *Structure* sorting out what he meant by "paradigm" (see Wray 2011, chapter 3). He came to realize, as numerous critics had suggested, that he had made a mess of this concept (see Kuhn 1997/2000, 298; see also Kuhn 1991/2000b, 221). He acknowledged that it was Masterman who made him realize that "a paradigm is what you use when the theory isn't there" (see Kuhn 1997/2000, 300). By the early to mid-1970s, Kuhn had clarified what *he* had meant, distinguishing (i) exemplars from (ii)

---

when I first read that. I was at Princeton, I went and put a note on the bulletin board of the department office, saying, 'I have just read the most exciting piece that I've read since I discovered Alexander [*sic*] Koyré!'" (Kuhn 1997/2000, 304).

theories, and (iii) disciplinary matrices. He insisted that the term "paradigm" should be applied only to exemplars. Theory changes he came to conceptualize as lexical changes of a very specific sort, involving a violation of what he called "the no-overlap principle" (Kuhn 1991/2000a, 92). And the disciplinary matrix was the broadest notion, encompassing theory, values, goals and practices. Despite these clarifications, many who discuss Kuhn's view still refer to changes of theory as paradigm changes. The phrase has stuck.

So, despite the concerns that early critics raised about the paradigm concept, the term has become thoroughly entrenched. In fact, the concept has been transformed from an analyst's term, used by philosophers, sociologists and historians of science, into an actor's term, used by working scientists. As we saw earlier, it was picked up, indeed, enthusiastically embraced, by social scientists. But natural scientists also now use it. One can find many scientific papers in leading journals using the term paradigm. For example, a recent article in *Science* studying perception and cognition using mice as subjects refers to "an operant conditioning behavioral paradigm," as well as a "stimulus paradigm" (see Neubarth et al. 2020). The extent to which the term is used to mean exemplar, as Kuhn wanted, is unclear, but its use is now quite commonplace in science.

In Kuhn's defense, Ian Hacking has recently argued that, in invoking the paradigm as exemplar concept, Kuhn was grappling with a problem that has vexed philosophers since Aristotle's time. Hacking dubs the problem the "paradeigmatic muddle" (see Hacking 2016, 99). The problem is to explain how we reason from examples, a topic Aristotle took up in his *Rhetoric*. Kuhn claims that scientists, working in a normal science research tradition, reason from examples. They appeal to the solution to one research problem in order to address another hitherto-unsolved research problem. For example, in Max Planck's effort to solve the black-body problem he appealed to the methods Boltzmann had developed in his research on the behavior of gases. Hacking suggests that the difficulties that Kuhn faced in his efforts to elucidate the "paradigm concept" are "pervasive in the family of notions we variously call analogy, models, similarity, likeness, resemblance" (Hacking 2016, 109). Though Hacking grants that "no one has given a satisfactory analysis of Kuhn's idea of a paradigm as example," he praises Kuhn "for giving new life to ... an ancient idea" (Hacking 2016, 109).

The fallout related to the charge of irrationality is more complex. Kuhn admitted that "of the things that surprised me tremendously in the reactions to *Structure*, a major one was the talk about irrationality, for that was something that had never occurred to me" (see Kuhn 1990/2016, 22). Addressing this concern became a major preoccupation for Kuhn for the rest of his life. He was never really able to convince many philosophers of science that his view did not impugn the rationality of science. And it was Lakatos' concern with the rationality of scientific revolutions that caused the most unease. For many philosophers of science, the alleged incommensurability between competing paradigms made it challenging to understand how a choice between competing theories could be rationally grounded. Indeed, Kuhn's claim that competing theories are incommensurable was often interpreted as a claim that competing theories are not even comparable. In fact, this reading of Kuhn persists. Robert DiSalle, for instance, argues that Kuhn's "account of paradigm shift as conversion ... represents scientists as incapable of mutual understanding" (DiSalle 2002, 194).

John Schuster, a former student and then one-time colleague of Kuhn's, argues that,

> from the time of being philosophically ambushed by Popper and the Lakatosians at the Bedford College Conference ... Kuhn began to retreat from creative thought about historical process, slowly sinking into a maelstrom of unending, unwinnable debate with philosophers about rationality and method. (Schuster 2018, 418)

It is fair to say Kuhn never won this debate, but it is also fair to say that his opponents were not winners either.

In fact, part of the problem was due to a misunderstanding. Kuhn was presenting a radically different conception of rationality than the conception of rationality that was widely accepted in philosophy of science in the early 1960s. Prior to Kuhn, the rationality of science was widely understood narrowly in terms of logic, and specifically of the logical relationship between evidence and theory. Science is rational, it was widely believed, insofar as scientific theories are accepted – or in Popper's case, rejected – in light of the evidence. Confirmation theory would provide us with the rules of logic that would tell us when it was and when it was not rational to accept a theory. And all scientists faced with the same evidence should accept the same theory. All scientists, after all, are bound by the same logical rules.

Kuhn, on the other hand, was working with a conception of rationality that permitted rational disagreement (see Kuhn 1969/2012, 198–199). Scientists confronted with the same evidence, he claimed, may not agree about which of two competing theories is superior. And this is not a failing. Further, Kuhn even insists that the differences between scientists, even subjective differences, are not an impediment to science. Rather, such differences could have, and did in fact have, a positive impact on science, as they were sources of novel ideas, methods and practices (see Kuhn 1970/2000, 134). Thus, according to Kuhn, the influence of subjective factors on scientists was not antithetical to scientific rationality. The locus of rationality had shifted, from a logical relationship between evidence and theory, to the research community's response or reaction to evidence (see Kuhn 1970/2000, 134).

Indeed, Kuhn regarded this shift to the level of the community as fundamental. He explains this at length in the Postscript to the second edition of *Structure*. After describing the sorts of scientific research communities he has in mind, and some techniques that might be used to identify them, Kuhn notes that "communities of this sort are the units that [*Structure*] has presented as the producers and validators of scientific knowledge" (Kuhn 1969/2012, 177). The techniques Kuhn describes are those that were becoming commonplace in the sociology of science and scientometrics, for example, analyses of "linkages among citations" (see Kuhn 1969/2012, 177). Such techniques, he claimed, could help us identify the networks and communities that constitute the subgroups that undergo revolutionary changes of theory.

Kuhn suggests that the traditional individualist approach to philosophy of science is deeply problematic.

> Groups ... should ... be regarded as the units which produce scientific knowledge. They could not, of course, function without individuals as members, but the very idea of scientific knowledge as a private product presents the same intrinsic problems as the notion of a private language. (Kuhn 1970/2000, 148)

Indeed, this is why Kuhn insists that there is a "sociological base" to his theory of scientific knowledge and change, something that was in the early 1970s, and, to some degree still is, antithetical to philosophers of science (see Kuhn 1970/2000, 148).[9] Specifically, Kuhn insists that in the development of a scientific field, the "transition to a closed specialists' group was part of the transition to maturity," which he believes is intimately connected with "the emergence of puzzle-solving" that characterizes the

---

[9] Hoyningen-Huene argues that "Kuhn uses two basic assumptions that import sociology into the philosophy of science ... that *communities* and not individuals should be seen as the basic agents of science ... [and] these communities must be characterized by the specific cognitive *values* to which they are committed" (Hoyningen-Huene 1992, 492; emphasis in original).

tradition-bound research associated with normal science (see Kuhn 1970/ 2000, 149).

This theme was developed further in much of Kuhn's work after *Structure*. In the Preface to *Essential Tension*, for example, Kuhn notes that

> traditional discussions of scientific method have sought a set of rules that would permit any *individual* who followed them to produce sound knowledge. I have tried to insist, instead, that, though science is practiced by individuals, scientific knowledge is intrinsically a *group* product and that neither its peculiar efficacy nor the manner in which it develops will be understood without reference to the special nature of the groups that produce it. (Kuhn 1977, xx)

Kuhn would continue to discuss this issue in the 1980s and early 1990s, especially in his papers "The Road since *Structure*" and "The Trouble with the Historical Philosophy of Science" (see Kuhn 1991/2000a and 1992/ 2000).

The philosophical community, though, has been slow in making this shift from the individual scientist to the community. Some, like Fred D'Agostino and Alexander Rueger, for example, have drawn attention to Kuhn's risk-spreading argument (see Rueger 1996; D'Agostino 2005). And it is a central theme in my earlier book on Kuhn (see Wray 2011). But the impact of Kuhn's insightful account of scientific rationality is still slight.

Further, significantly, and contrary to what his critics seems to suggest, Kuhn did not say that logic did not matter. In that respect, the charge of irrationality is wholly misguided. Indeed, Kuhn is explicit that he believes that when scientists make a choice between competing theories, they have "good reasons for choosing between them" (see Kuhn 1970/2000, 127; see also 156). His concern was that the sorts of reasons that persuade scientists do not reach the level of proof that we find in mathematics, which commands assent (see Kuhn 1970/2000, 156). That is why Kuhn was insistent that rational scientists can disagree.

Further, Kuhn was insistent that there is no theory-neutral vocabulary that would allow scientists to make a "point-by-point comparison of two successive theories," one in which "the empirical consequences of both [theories] can be translated without loss or change" (see Kuhn 1970/2000, 162). Rather over-optimistically, in 1970 Kuhn claimed "philosophers have now abandoned hope of achieving any such ideal," for "in the transition from one theory to the next words change their meanings or conditions of applicability in subtle ways" (Kuhn 1970/2000, 162–163). This appeal to meaning incommensurability would prove to be a sticking point with many philosophers. Indeed, philosophers continue to debate the merits and plausibility of Kuhn's incommensurability theses (see, for example, the various papers in Hoyningen-Huene and Sankey 2001).

There were other consequences of the rationality debate. Lakatos' critique of Kuhn, insofar as it was aimed at Popper as well, contributed to the downfall of Popper and Popperianism. Indeed, by 1965, it seems that Lakatos and Feyerabend were conspiring against Popper, and were set on changing the philosophical agenda he had set (on Feyerabend, see Collodel 2016). And it seems they were quite successful, at least from a rhetorical point of view. Soon thereafter, it would be quite commonplace for Popper to be discussed in a manner that suggested that he was a Logical Positivist. That is, he was indiscriminately grouped with Logical Positivists, as one concerned with the logic of science, rather than being attuned to the importance of the history of science for the philosophy of science, as the younger generation of philosophers of science were. Popper's falsificationism was less often regarded as an interesting position in itself, and more often as merely a variant of the sort of thing the Logical Positivists had been doing for decades.

Kuhn, though, seemed to regard Popper as an ally against the Logical Positivists' conception of science. More precisely, he saw that they both had developed views that were similar in a number of respects and that went against key tenets of Logical Positivism. In his replies at the Bedford College conference, Kuhn notes that:

> on almost all the occasions when we turn explicitly to the same problems, Sir Karl's view of science and my own are very nearly identical. [1] We are both concerned with the dynamic process by which scientific knowledge is acquired rather than with the logical structure of the products of scientific research. Given that concern, [2] both of us emphasize, as legitimate data, the facts and also the spirit of actual scientific life, and both of us turn often to history to find them. From this pool of shared data, we draw many of the same conclusions. [3] Both of us reject the view that science progresses by accretion; both emphasize instead the revolutionary process by which an older theory is rejected and replaced by an incompatible new one; and both deeply underscore the role played in this process by the occasional failure of the older theory to meet challenges posed by logic, experiment, or observation. Finally, Sir Karl and I are united in opposition to a number of the most characteristic theses of classical positivism. [4] We both emphasize, for example, the intimate and inevitable entanglement of scientific observation with scientific theory; [5] we are correspondingly skeptical of efforts to produce any neutral observation language; and [6] we both insist that scientists may properly aim to invent theories that *explain* observed

phenomena and that do so in terms of *real* objects, whatever the latter phrase
may mean. (Kuhn 1970/1977, 267; emphasis in original; numerals added)

Though, as Kuhn notes, there are many similarities between his views and
those of Popper, I suspect that Kuhn is being conciliatory here, perhaps
excessively so. He was probably motivated, in part, by the desire to avoid an
excessively aggressive confrontation with Popper, who had a reputation for
being surly and aggressive (Edmonds and Eidinow 2002, 169 and 176; see
also Edmonds 2020, 245). Popper's personality was notorious. In fact, as
David Edmonds notes, Moritz "Schlick thought [Popper] far too rude to
receive an invitation" to the Vienna Circle meetings (see Edmonds 2020,
22). Indeed, Kuhn makes clear at the beginning of his Bedford College
conference paper that he is "not so sanguine as Sir Karl about the utility of
confrontations" (Kuhn 1970/1977, 266).[10]

Kuhn, though, also makes clear that his view differs from Popper's view
in fundamental ways. Specifically, he claims that, "to turn Sir Karl's view
on its head, it is precisely the abandonment of critical discourse that marks
the transition to a science" (Kuhn 1970/1977, 273). Thus, despite the long
list of similarities that Kuhn identified between their views, Kuhn rejected
what was most central to Popper's critical rationalism.[11] Popper insisted
that critical discourse was the mark of rationality. Kuhn, on the other
hand, was suggesting that what was most characteristic of the natural
sciences and accounted for their impressive successes, specifically, the
sorts of successes that we tend to associate with scientific progress, is the
fact that they set aside critical discussions about the fundamentals, at least
during periods of normal science.

It seems that not everyone has been persuaded by Popper regarding
Kuhn's view of the role and function of dogmatism in normal science. In
a paper, published in 2002, dedicated to reexamining the Popper–Kuhn
controversy, John Worrall suggests that he has grasped Kuhn's point.
Worrall argues that "Kuhn, contrary to Popper's interpretation . . . should
be seen *not* as advocating dogmatism, but rather as advertising the fact that
'commitment' to the sort of framework supplied by well-developed science

---

[10] John Heilbron recounts an earlier encounter between Kuhn and Popper, in 1962 at Kuhn's home,
which makes clear how mismatched they were. "The guest was allergic to smoke and the host was
addicted to cigarettes. Communication across worldviews, never an easy matter, failed altogether
owing to coughing fits on the one side and tobacco fits on the other" (see Heilbron 1998, 510).

[11] Kuhn also sees that in an important sense, Popper's critical rationalism is an extension of the Logical
Positivists' program. In his response to his critics at the Bedford College conference, Kuhn notes that
"for Sir Karl and his school, no less than for Carnap and Reichenbach, canons of rationality . . .
derive exclusively from those of logical and linguistic syntax" (Kuhn 1970/2000, 227).

brings enormous epistemic benefits" (Worrall 2002, 81). Thus, read charitably, much of what Kuhn claims is quite sensible and not especially contentious. Even claims that his critics suggest are ludicrous are often sensible when seen in a sympathetic light.

The fallout related to the charge of relativism is also interesting. Cast as a relativist, Kuhn became a convenient foil for many. And this criticism seems to have stuck. No doubt, the Strong Programme's association with Kuhn, and the fact that they explicitly identify as relativists, further underscored this reading of Kuhn, that is, for those who were inclined to read him that way in the first place.

Significantly, Kuhn did try to address this concern. But his early remarks in response to this charge were largely ignored. In his replies to his critics he makes clear that "one scientific theory is not as good as another for doing what scientists normally do," adding, "in that sense *I am not a relativist*" (see Kuhn 1970/2000, 160; emphasis added). He believed without a doubt that science progresses, and the theories developed and accepted more recently in the history of a scientific field are superior to those developed earlier in the history of the field. He invokes an interesting and insightful picture to make his point, a picture that would loom large in his later writings.

> Imagine ... an evolutionary tree representing the development of the scientific specialties from their common origin in ... natural philosophy. Imagine, in addition, a line drawn up that tree from the base of the trunk to the tip of some limb ... Any two theories found along this line are related to each other by descent. Now consider two such theories, each chosen from a point not too near its origin. I believe it would be easy to design a set of criteria – including maximum accuracy of predictions, degree of specialization, number (but not scope) of concrete problem solutions – which would enable any observer involved with neither theory to tell which was older, which the descendant. For me, therefore, scientific development is, like biological evolution, unidirectional and irreversible. (Kuhn 1970/2000, 160).

The unequivocal superiority of more recently developed theories, Kuhn insists, should make clear that he is no relativist. A relativist, he suggests, would maintain that there is no basis for claiming one theory is superior to another.

Kuhn suspected the reason that he is often cast as a relativist is that he denied that scientists can "compare theories as representations of nature, as statements about 'what is out there'" (Kuhn 1970/2000, 160). He even denies that we can effectively make judgments to the effect that a theory developed more recently in a field "is a better approximation to the truth" (see Kuhn 1970/2000, 160).[12] But, he also insists that an explanation of the progress of science does not depend on such a judgment (see Kuhn 1970/ 2000, 161).

Papineau's version of the charge of relativism raises different issues. Papineau suggests that relativists like Kuhn and Feyerabend believe that observational evidence is infected by a scientist's theoretical presuppositions. The term "infection" implies a serious problem, specifically, that observational evidence is no longer objective. On the one hand, Papineau is correct to claim that Kuhn believes that there are no theory-independent observations. But Kuhn would not have described this as a case of infection. Rather, as far as Kuhn was concerned, effective observations in science require the aid of theory and theoretical presuppositions. For example, chemists seemed unable to see certain chemical data as evidence of chemical reactions until they looked at the data with knowledge of Dalton's law of fixed proportions (see Kuhn 1962/2012, 132–133). Though theoretical presuppositions may lead scientists to be partial in what aspects of the world they attend to, the data are not aptly described as theory-infected, insofar as that term implies a distortion of reality. Sometimes the presuppositions merely aid scientists by directing their attention to what is salient, given their research interests.

Again, charitable readers of *Structure* did not see the problems his most vociferous critics saw. Gerald Doppelt, for example, even offered a defense of Kuhn's epistemological relativism (see Doppelt 1978). The key is to get clear on the sense in which Kuhn is aptly characterized as a relativist.

Though philosophers often associated Kuhn's view with some form of relativism of a problematic nature, the sociologist of science Robert Merton could not understand the basis of this criticism against Kuhn. In a letter from 1975, in which Merton is commenting on a draft of Kuhn's paper "Mathematical versus Experimental Traditions in the Development of Physical Science" (see Kuhn 1976/1977b), Merton singles out for special attention the following passage from Kuhn's manuscript.

> If ... the ACCUMULATION of concrete and APPARENTLY PERMANENT PROBLEM SOLUTIONS is a measure of SCIENTIFIC

---

[12] Kuhn has in mind here Popper's work on verisimilitude (see Kuhn 1970/2000, 161). Popper argued that one theory is closer to the truth than another if either (i) it contains fewer falsehoods and at least as many truths as the other, or (ii) if it contains more truths and no more falsehoods than the other. By the mid-1970s, Popper's account of verisimilitude had been shown to be deeply flawed by both David Miller and Pavel Tichý (see Psillos 1999, chapter 11). Indeed, the notion of approximate truth has defied explication, which has proved somewhat problematic for scientific realists.

> PROGRESS, these fields [that is, the mathematical sciences] are the only
> parts of what was to become the physical sciences in which
> UNEQUIVOCAL PROGRESS WAS MADE during antiquity. (Merton
> 1975; the emphases are Merton's).

Merton sees here both (i) a clear statement of what Kuhn means by
scientific progress, and (ii) a clear statement that Kuhn does not deny that
science progresses. That is, Merton sees in this passage remarks that are
irreconcilable with the sort of relativist reading of Kuhn that was becoming
quite widespread among philosophers. Merton continues:

> I have been arguing all along, since at least the late 1950s, let alone the
> marvelous year of 1962, TSK has steadfastly rejected that kind of (simple-
> minded) relativism which would seriously declare that there is no selective
> accumulation of scientific knowledge, that pushpin is as good as poetry, &c.
> &c. That whether it is described as "scientific progress," "the growth of
> scientific knowledge". . . &c., such growth can be historically demonstrated
> and epistemologically described without accepting the (simpleminded)
> notion of "unilinear, inexorable, steady accelerating accumulation of know-
> ledge." (Merton 1975)

Merton rightly saw that Kuhn did not think that scientific progress was
linked to such a linear, steady accumulation of knowledge.

The fallout related to the concerns about Kuhn's aims, that is, whether
they were normative or merely descriptive, would affect the philosophy of
science profoundly, and continues to do so today. These concerns have
given rise to a number of debates about methodology in the philosophy of
science. Kuhn himself subsequently contributed to the debate about the
relationship between the history of science and philosophy of science, and
whether there is in fact an integrated field, HPS. Kuhn "[urged] that
history and philosophy of science continue as separate disciplines. What
is needed is less likely to be produced by a marriage than by active
discourse" (Kuhn 1976/1977a, 20). This debate continues (see, for example,
Giere 1973; McMullin 1976; Burian 1977; Scholl 2018).

There is also a related debate about the value of case studies from the
history of science for the philosophy of science. The key question in this
debate is: what sort of general philosophical claims can be drawn from
a case study in the history of science? As part of the Historical School in the
philosophy of science, Kuhn made popular appeals to case studies. But it is
far from clear that one, or even a few cases studies, can support the sorts of
general claims that philosophers aim to make about science. These debates

are still not resolved and continue on (see, for example, Pitt 2001 and Burian 2001; and more recently, Sauer and Scholl 2016).

Recently, Moti Mizrahi has collected data on the use of case studies in key philosophy of science journals, specifically *Philosophy of Science*, *British Journal for the Philosophy of Science* and two other journals. Though there has been a steady increase of articles using case studies in the two most important philosophy of science journals since their inception, there is no evidence that Kuhn's *Structure*, in particular, had an impact (see Mizrahi 2020, figures 3 and 4). That is, there is no marked increase in the use of case studies in articles published in these journals on or shortly after 1962, 1970 or 1996, when the first, second and third editions of *Structure* were published. So, this development may be independent of Kuhn's contribution.

But to the question – Is Kuhn offering a normative account of science, an account that is continuous with the traditional aims of philosophy of science, or a descriptive account? – Kuhn has a straightforward reply. He claims his remarks about science should be read as both descriptions *and* prescriptions. As he explains, "if I have a theory of how and why science works, it must necessarily have implications for the way in which scientists should behave if their enterprise is to flourish" (Kuhn 1970/2000, 130). Indeed, I think that these days, now that philosophy of science is thoroughly naturalized, most philosophers of science would be satisfied with Kuhn's answer. When one has an account of the factors that contribute to the success of science, one has compelling grounds for normative claims to the effect that scientists should continue in these successful practices.

In summary, I have argued that the Kuhnian view became rather settled in the philosophy of science quite early, and philosophers of science were content to have the Kuhnian view as a convenient foil. Thus, it is not uncommon to see philosophers addressing Kuhn's view, even quite recently, focusing narrowly on *Structure*, as if Kuhn had had nothing to say afterwards.

It seems that Kuhn contributed to this situation to some extent. And he even seems to acknowledge this. Kuhn notes:

> people treated me as though I were a fool! I want to say, how the hell could anybody ever have thought that I would believe anything like that! That really was fairly destructive, and I fairly early simply stopped reading the things about me, from philosophers in particular. (Kuhn 1997/2000, 315)

His failure to address his philosophical critics, though understandable, allowed certain criticisms to stand.

In closing, it is worth contrasting these early influential criticisms of Kuhn's *Structure* with another early assessment of the book. Hesse reviewed *Structure* in the journal *Isis*. According to Hesse,

> this is . . . the kind of book one closes with the feeling that once it has been said, all that has been said is obvious, because the author has assembled from various quarters truisms which previously did not quite fit and exhibited them in a new pattern in terms of which our whole image of science is transformed. (Hesse 1963, 286)

Juxtaposing Hesse's assessment of the book with the criticisms discussed above, it is no wonder that Kuhn was "tempted to posit the existence of two Thomas Kuhns" (Kuhn 1970/2000, 123). At the Bedford College conference he felt that only Margaret Masterman had read Kuhn$_1$, the same book he himself was familiar with. Popper, Feyerabend, Lakatos, Toulmin and Watkins, on the other hand, had read Kuhn$_2$, a superficially similar, but obviously profoundly different book (see Kuhn 1970/2000, 124).

As Kuhn aged, he seems to have softened and developed a sense of humor about it all. Thus, at a conference in the 1989 he remarked: "I do, in short, really believe some – though by no means all – of the nonsense attributed to me" (Kuhn 1991/2000b, 220). Here, he specifically had in mind his commitment to claims such as the following: "the heavens of Greeks were irreducibly different from ours" (Kuhn 1991/2000b, 220).

CHAPTER 10

Kuhn and the Contemporary Realism/ Anti-Realism Debates

The impact that Thomas Kuhn and *The Structure of Scientific Revolutions* had in the philosophy of science was not always in the most obvious places, and not necessarily where he intended to have an impact. As mentioned in Chapter 2, one of Kuhn's legacies to the philosophy of science is the problem of theory change. In this chapter I want to examine how the problem of theory change has influenced the realism/ anti-realism debates.

Kuhn was not a key player in the contemporary realism/anti-realism debates in philosophy of science, that is, the debates that gained momentum around 1980 or so, with the publication of Bas van Fraassen's *Scientific Image* and Larry Laudan's "Confutation of Convergent Realism" (see van Fraassen 1980 and Laudan 1981; also, see Wray 2018a). But Kuhn had a significant influence on these debates. He played an important role in focusing philosophers' attention on the problems raised by *changes of theory*. In contrast, the focus of the realism/anti-realism debates of the 1950s and 1960s, which predated the publication of *Structure*, was on *the meaning of theoretical terms*. Thus, the particular shape of the contemporary realism/anti-realism debates owes something to the publication of *Structure*.

I begin by outlining the realism/anti-realism debates of the 1950s and 1960s. I trace these early debates to the research program of the Logical Positivists. I then briefly discuss the fate of instrumentalism, the realists' foil in the debates of the 1950s and 1960s, as well as the fate of Karl Popper, as he was an active participant in the early debates, but not in the later ones. Then I show how the realism/anti-realism debates changed in response to Kuhn's *Structure*. Finally, I analyze both (i) remarks Kuhn made in his later publications that have some relevance to realism and anti-realism, and (ii) how other scholars have situated Kuhn in the contemporary debates. Kuhn's real impact on those debates, I argue, is the problem that he presented philosophers with, that is, the problem of theory change.

The Realism/Anti-Realism Debates, Before Kuhn

Up until the mid-1960s, many of those involved in the realism/anti-realism debates in philosophy of science focused on understanding the meaning of our theoretical language in science, specifically, the meaning of terms that refer to unobservable entities, like electrons. These are the realism/anti-realism debates to which Popper, J. J. C. Smart and Grover Maxwell contributed. Paul Feyerabend, Rudolf Carnap and Herbert Feigl also contributed to these debates (see, for example, Feyerabend 1958/1981; 1960/1981; and 1964/1981; see Carnap 1956).

The realists involved in these debates had a very specific opponent in mind. Their opponents were *instrumentalists*. And these realists were especially concerned with *physicists*, rather than philosophers. Instrumentalism was regarded as a view widely held by physicists, specifically Niels Bohr and the others associated with the Copenhagen Interpretation of quantum mechanics.

Smart, for example, claimed that he was "concerned with the ontological status of the physicist's theoretical entities" (Smart 1963, 27). He identified his adversaries as those who "say that [the subatomic entities of physical theory] are not so much part of the furniture of the world as *useful conceptual devices* for predicting the behaviour of macroscopic objects such as stones and galvanometers" (Smart 1963, 16; emphasis added). According to Smart, "the most extreme form of this view which I wish to oppose is that sentences about electrons, protons, and the like can be *translated* into sentences about galvanometers, cloud chambers, and the like" (Smart 1963, 27; emphasis in original). Smart claims that "this is the philosophy of science made popular by Ernst Mach and adopted by many physicists when they branch off into philosophy" (see Smart 1963, 16).[1] He also suggests it is "the philosophy which was originally behind ... Heisenberg's matrix mechanics ... and the so-called 'Copenhagen interpretation' of quantum mechanics" (see Smart 1963, 16–17). In contrast to this view, Smart argues that the subatomic entities of physical theory really are "part of the furniture of the world." Electrons and protons really are as our theories describe them.

Maxwell had the same opponents in mind in his often-discussed paper, "The Ontological Status of Theoretical Entities" (see Maxwell

---

[1] Mach, for example, claims that "the atomic theory ... is a mathematical *model* for facilitating the mental production of facts" (Mach 1893/1960, 589). Similarly, he notes that "*all* hypothesis [*sic*] formed for the explanation of new phenomena" are mental expedients that "have nothing whatever to do with the phenomena *itself*" (Mach 1893/1960, 590).

1962).[2] As Maxwell explains, he is taking issue with the following views: "[i] that the entities referred to by scientific theories are only convenient fictions, . . . [ii] that talk about such entities is translatable without remainder into talk about sense contents or everyday physical objects, . . . [and] [iii] that such talk should be regarded as belonging to a mere calculating device and, thus, without cognitive content" (Maxwell 1962, 3; numerals added). Maxwell claims that he had initially believed that no one *really* took such views seriously. But then he realized that a number of "the instrumentalist views of outstanding physicists such as Bohr and Heisenberg" made it necessary for him to take these views seriously, and show why they were flawed (see Maxwell 1962, 3). Maxwell's argument that the observables and the theoretical are not categorically distinct, but lie on a continuum, aimed to show that there is no *ontological* significance to something being observable rather than unobservable, contrary to what he took Bohr and other physicists to be claiming. Hence, instrumentalism was not a viable position.

Popper, like Maxwell, was concerned with the pernicious effects of instrumentalism, and Bohr's instrumentalism was his target as well (see Popper 1956/1963, § 2). Popper was concerned that instrumentalism had "become an accepted dogma. It may well now be called the 'official view' of physical theory since it is accepted by most of our leading theorists of physics (although neither by Einstein nor by Schrödinger)" (see Popper 1956/1963, 134).[3]

As we see, the key players in the realism/anti-realism debates of the 1950s and 1960s contrasted their view with some form of instrumentalism. Whether or not any *philosopher of science* at that time actually held this position is unclear. It is doubtful that many did. In his 1961 book, *The Structure of Science*, Ernest Nagel critically analyzes the instrumentalist view of science. But, with the exception of one person, Stephen Toulmin, those whom Nagel discusses were either not alive or not philosophers.[4] In

---

[2] The continuing significance of Maxwell's paper owes something to the fact that Bas van Fraassen chose it as a foil against which to present his own constructive empiricism in *The Scientific Image* (see van Fraassen 1980, 14–18). It is anthologized in widely used textbooks in philosophy of science, including Kourany (1998) and Curd et al. (2013).

[3] When Popper wrote *The Logic of Scientific Discovery* (*LSD*), one of his principal foils was *conventionalism* (see Popper 1935/1992, Chapter 4, § 19–20). But his criticism of conventionalism in the 1930s was more or less the same as his criticism of instrumentalism in the late 1950s. Thus, in *LSD*, he accused the conventionalists of employing strategies that make theories less falsifiable, revising them ad hoc in light of otherwise falsifying observations. He identified Pierre Duhem and Henri Poincaré as conventionalists.

[4] Nagel identifies the following thinkers as instrumentalists: C. S. Peirce, Frank Ramsey, Moritz Schlick, John Dewey, W. H. Watson, Gilbert Ryle and Stephen Toulmin (see Nagel 1961, 129, fn. 22). All of the

these early debates, physicists in particular were alleged to be instrumentalists. Thus, not surprisingly, these early debates were principally focused on *physics*. Whether the concerns of the realists affected or were relevant to other sciences was never discussed. Indeed, the preoccupation with physics may explain why the topic of reductionism also figured in these early debates.[5]

Incidentally, it is doubtful that these philosophers attacking instrumentalism understood Bohr's view. For example, contrary to what Popper suggests, Bohr did not appeal to the principle of complementarity to save quantum theory from falsification (see Bohr 1948). Further, it is questionable whether it is correct to regard the Copenhagen Interpretation as a form of instrumentalism. Bohr was concerned to show that the conception of reality that classical mechanics presupposed, which is tied to what he calls "the principle of causality," is fundamentally at odds with quantum mechanics (see Bohr 1948, 313). He was not suggesting that the entities posited by physical theories did not really exist. Indeed, Don Howard has persuasively argued that what is commonly called the "Copenhagen Interpretation" does not represent Bohr's view. The term, as well as the view associated with it, was created by Werner Heisenberg in the mid-1950s (see Howard 2004).

One of the key questions for those engaged in the debates in the late 1950s and early 1960s was whether an instrumentalist interpretation of theoretical terms was defensible, or whether such terms should be interpreted literally, as realists urge (see Smart 1963, 39; Maxwell 1962, 3). This debate was about the *meaning* of theoretical claims in science. Indeed, that the debate was framed in this way is not a surprise when one remembers the central role that philosophy of language played in philosophy in the 1950s and 1960s.

These debates focusing on the meaning of theoretical terms had their origins in the research program of the Logical Positivists. The Vienna Circle positivists were concerned about the status of metaphysical claims and their pernicious effects on science (see, for example, Carnap 1932/1959). Their verification principle was intended to be a means for distinguishing between meaningful discourse and meaningless, maybe even dangerous, types of discourse (see Schlick 1932–1933/1959 § 2; also Popper 1935/1992, 12).

---

sources by these authors that Nagel cites are from before 1954. I learned that Watson was a physicist from the subtitle of his 1963 book, *Understanding Physics Today: A Physicist's Quest for an Intuitive Grasp of Atomic Essence.*

[5] I thank Nancy Cartwright for reminding me that these same philosophers were also concerned with the issue of reductionism.

Given the anti-metaphysical attitude of the Vienna Circle, it is not surprising that they were inclined toward anti-realism. But by the 1950s and early 1960s the successors of the Vienna Circle, its heirs and critics – Maxwell, Smart and Popper – were realists.

Interestingly, in the early 1930s Moritz Schlick wrote a paper on positivism and realism in an attempt to distance the positivists' view from idealism, a view that is sometimes associated with anti-realism (Schlick 1932–1933/1959). Though Schlick implies that realism and positivism are compatible, we should not be misled into thinking that he was a realist in the sense in which either Smart or Popper or many contemporary scientific realists are realists. As Schlick explains, his aim is merely to make clear that "consistent empiricism does *not* deny the existence of the external world" (Schlick 1932–1933/1959, 107). This was not the issue dividing realists and instrumentalists in the early 1960s, nor is it an issue that divides contemporary realists and anti-realists.

In true Logical Empiricist fashion, Schlick explains that "the denial of the existence of a transcendental external world would be just as much a metaphysical statement as its affirmation. Hence, the consistent empiricist does not deny the transcendental world, but shows that both its denial and affirmation are meaningless" (Schlick 1932–33/1959, 107).[6] Schlick had no concern with (i) the issues that concerned those involved in the realism/anti-realism debate of the 1950s and 1960s, or (ii) with external-world skepticism.

Indeed, even in the 1940s Philipp Frank, another member of the Vienna Circle, was citing Percy Bridgman approvingly, to explain the positivists' reaction to Albert Einstein's general theory of relativity (see Frank 1949/1961, 31, fn. 8). Bridgman's operationalism is often characterized as a form or close relative of instrumentalism, and certainly always as a form of anti-realism (see, for example, Popper 1935/1992, 386; Popper 1963, 82–83; and Chang and Cartwright 2014, 411–413).

Thus, the Vienna Circle's focus on the language of science set the stage for the realism/anti-realism debates of the late 1950s and the early 1960s.

The Fate of Instrumentalism and Popper

Before turning to a discussion of the impact of *Structure* on the realism/anti-realism debates, I want to briefly consider the fate of the older debates, the debates between realists and instrumentalists.

---

[6] Apparently, according to Herbert Feigl, Schlick moved closer to anti-realism under the influence of Carnap and Wittgenstein, after his move to Vienna in the 1920s (see Neuber 2011, 167).

Generally, those involved in the contemporary debates who identify as anti-realists do not identify as instrumentalists. Van Fraassen, for example, claims that theories are to be interpreted literally. That is, he believes that claims attributing properties to unobservables are either true or false (see van Fraassen 1980, 9–13). In this respect, van Fraassen's anti-realism is not the sort of anti-realism that Smart or Maxwell were criticizing. What motivates van Fraassen's anti-realism is that the available data do not provide adequate grounds for believing the claims our theories make about unobservables. He never suggests that such claims are reducible to claims about observables. This is a very different sort of anti-realism than the anti-realism associated with the classic instrumentalist view of theories discussed in the late 1950s and early 1960s. Instrumentalists, we were told, insist that theories are to be evaluated in terms of their usefulness (see Popper 1956/1963, 151). So, as such, an instrumentalist would say that claims about unobservables are not aptly characterized as either true or false.[7]

Some contemporary anti-realists are pragmatists of sorts, and pragmatism has sometimes been associated with instrumentalism. Consider, for example, Larry Laudan's focus on the problem-solving abilities of scientific theories (see Laudan 1977). But it would be a mistake to regard Laudan as an instrumentalist. He does not suggest that we should attempt to reduce claims about unobservables to claims about observables, as the instrumentalists of the 1950s and 1960s did, or allegedly did. Consequently, the pragmatism in the contemporary debates should not be equated with or reduced to some form of instrumentalism.[8] Thus, despite the differences among the anti-realists involved in the contemporary debates, generally they are not instrumentalists.[9] This is not especially surprising, given that instrumentalism was not a popular view among *philosophers of science* even in the earlier debates. Recall, as we saw earlier, that if anyone held such a view, it was scientists, and in particular physicists.

---

[7] Alan McMichael argues that van Fraassen is an instrumentalist. According to McMichael, van Fraassen is an instrumentalist in virtue of the fact that he believes that "statements of a scientific theory that deal with unobservable entities should be neither believed nor disbelieved. Their value is purely instrumental" (see McMichael 1985, 258). But van Fraassen is quite explicit that his view is not a form of instrumentalism (see van Fraassen 1980, 10).

[8] As a matter of historical fact, William James, a classic early pragmatist, was very sympathetic to Ernst Mach's instrumentalism (see, for example, James 1907/1949, 57). But in principle, instrumentalism and pragmatism are distinct and consequently separable.

[9] Recently, instrumentalism has gained some attention, as both Kyle Stanford and Darrell Rowbottom have attempted to revive it in some form (see Stanford 2006; Rowbottom 2019).

Let us now consider the fate of Popper in the realism/anti-realism debates. Popper was on the realist side in the earlier debates, and he was quite critical of instrumentalism, singling out Bohr and Heisenberg as paradigmatic instrumentalists (see, for example, Popper 1956/1963, 135–136; 149–153). From Popper's perspective, these instrumentalists made the quantum theory less falsifiable. Such a practice, Popper argues, is antithetical to the epistemic aims of science.

Popper, though, has not played a significant role in the contemporary realism/anti-realism debates. This is so for two reasons. First, Popper's criticisms of instrumentalism were rendered irrelevant, given the focus of the contemporary debates. From the 1970s to the early 2000s, instrumentalism was no longer the appropriate foil for those committed to defending realism. Second, Popper's brand of realism is difficult to place in the contemporary debates. This is so for two reasons. Most importantly, Popper did not think that scientists are in the business of proving that hypotheses and theories are true. He thought that testing was merely a means for determining which hypotheses or theories are *false*, and thus need to be discarded. He was insistent that theories always retain their hypothetical character (see Popper 1963, 72). This attitude toward theories is incompatible with the contemporary realists' claim that we have good reason to believe that our current theories are approximately true with respect to what they say about unobservable entities and processes.[10] The realists' arguments from the success of science, like the popular No Miracles Argument, are designed to give us reason to believe our current best theories are true or at least approximately true. Though not all realists in the contemporary debates make such an inference, it has played a significant role in many of the contemporary debates. Further, Popper's brand of realism was more concerned with the *attitude* that scientists should have if they are ever going to contribute to the advancement of science. He was insistent that instrumentalists would settle for theories that were merely instrumentally successful (see Popper 1956/1963; also Wray 2015b). But, according to Popper, such a threshold was inimical to genuine progress in science. A realist attitude, Popper claims, would spur scientists to test their theories vigorously, and thus aid them in the pursuit of true theories, even if they could never know if their theories were in fact true (see also

---

[10] I was reminded by one of the referees that there are *some* contemporary realists who are also Popperians, or at least influenced by Popper, for example, Alan Musgrave and Howard Sankey. But this is a minority position in the contemporary debates.

Feyerabend 1964/1981).[11] This motivational dimension of realism has also not played a large part in the current debates (for an exception, see Forbes 2017). Thus, it is not surprising that Popper is not a player in the contemporary debates. Popper's concerns, like instrumentalism, have been left behind. The post-Kuhnian realism/anti-realism debates focus on other issues.

But it would be a mistake to think that Popper left no trace on the current realism/anti-realism debates. His analysis of verisimilitude, that is, relative closeness to the truth, has contributed significantly to current analyses of the notion "approximate truth." This is so despite the fact that Popper's own analysis is recognized as fundamentally flawed (see Psillos 1999, 261–264).

Structure and the Problem of Scientific Revolutions

As I argued earlier, in Chapter 2, when Kuhn published *Structure* he made theory change a philosophical problem. Of course, changes of theory were widely acknowledged to exist before Kuhn. In fact, the changes of theory in twentieth-century physics had played a crucial role in the thinking of the members of the Vienna Circle positivists (see Frank 1949/1961, 30–33). Their epistemic modesty, including their aversion to metaphysics, was, in part, a consequence of what they learned from the then-recent developments in physics.[12]

But Kuhn studied theory change in a different manner than the Logical Positivists, and he was thus led to attribute a different sort of significance to theory change than the Logical Positivists attributed to it. Rather than focusing narrowly on the changes that occurred in twentieth-century physics, Kuhn examined a number of cases of theory changes in detail. Further, his aim was to determine what changes of theory can tell us about the growth of scientific knowledge and scientific progress.

---

[11] Of course van Fraassen also discusses what attitude one might take insofar as acceptance is an attitude (see van Fraassen 1980, 12). But whereas Popper claims that the attitude scientists take affects their prospects of getting at the truth, Van Fraassen makes no such assumption. Rather, he claims that acceptance of a theory, unlike belief in a theory, makes one less prone to having false beliefs.

[12] The Vienna Circle positivists were also influenced by the development of non-Euclidean geometries. Those developments were regarded as just as revolutionary as the changes in physics ushered in by Einstein and his contemporaries (see Frank 1949/1961, § 6 and § 10). Popper was also influenced by the revolutionary changes in physics in the early twentieth century. But the lessons he drew were that (i) we could never prove a theory is true through testing, and (ii) a proper scientific theory is always vulnerable to being falsified (see Popper 1974/1992, 37–39). Still, Popper remained a committed realist.

Kuhn came to believe that it is a mistake to think of successive theories in a field as getting ever closer to the truth (see, especially, Kuhn 1962/2012, 169–172). He thus argued that theory change was incompatible with the popular picture of scientific progress as cumulative (see Kuhn 1962/2012, 7). In making this connection, Kuhn made theory change an epistemic problem, and a challenge that scientific realists needed to address. In Ian Hacking's words, "the thought of a scientific revolution is not Kuhn's ... but Kuhn invites the idea that every normal science has the seeds of its own destruction. Here is an idea of perpetual revolution" (Hacking 1983, 8–9).[13]

Kuhn's claim that successive theories in a field are incommensurable further heightened the problematic nature of theory change. Not only was he suggesting that there was some loss through changes of theory, what is now sometimes called "Kuhn-loss," but he was also suggesting that the evaluation of competing theories was not a straightforward affair. This is because proponents of competing theories (i) do not agree about the standards by which theories should be evaluated and (ii) are often not able to communicate fully and effectively with each other. This is what he had in mind when he claimed that competing theories were incommensurable. But a number of Kuhn's critics took him to be claiming that competing theories could not even be *compared*, a claim that some regarded as a *reductio ad absurdum* of Kuhn's position (see, for example, Scheffler 1967, 82).

Kuhn's analysis of scientific revolutions really caused a stir and set philosophy of science in a new direction, including by giving new life to the realism/anti-realism debates. But the impact of his work on the realism/anti-realism debates was indirect, and not necessarily, in fact, seldom, cast in the distinctively Kuhnian terms that made *Structure* so popular. Nonetheless, that impact is undeniable.

To a large extent, the various pessimistic inductions that have played such a central role in the contemporary realism/anti-realism debates were a response to Kuhn's problem of theory change (see Hesse 1976; Putnam 1978; and Laudan 1981). Hilary Putnam is most explicit in connecting the pessimistic meta-induction, as he calls it, to Kuhn's philosophy of science. In *Meaning and the Moral Sciences* where Putnam introduces the

---

[13] Hacking's choice of terms is a bit unfortunate. The idea of perpetual revolution would come to be associated with Popper, who was critical of normal science, thinking that it involved the sort of dogmatism that he felt was antithetical to science, and with his own critical rationalism. What Hacking means is that, on Kuhn's view, there is no end to the cycle of normal science-crisis-revolution.

Pessimistic Induction, he asks: "How would our notions of *truth* and *reference* be affected if we decide *there is no* convergence in knowledge?" (Putnam 1978, 22; emphasis in original). He then proceeds to note that "this is already the situation according to someone like *Kuhn*, who is sceptical about convergence and who writes . . . as if the same term *cannot* have the same referent in different paradigms" (Putnam 1978, 22; emphasis added). Putnam overstates things here. Kuhn would say that the same term, for example, atom or molecule, *may* not have the same referent in competing paradigms, not that it *cannot* have the same referent.

In its most general form, the Pessimistic Induction suggests that the fate of our currently accepted theories in science is not likely to be any different from the fate of the theories they replaced. Like our current best theories, these earlier theories were predictively and explanatorily successful. In these important respects, our current theories are indistinguishable from their predecessors. Because the currently accepted theories seem to have no relevant feature that distinguishes them from the theories they replaced, it seems that we are not warranted in believing they are immune from being replaced in the future (see Wray 2015a). Addressing the Pessimistic Induction is widely regarded as a key issue in the contemporary realism/ anti-realism debates (see, for example, Psillos 1999; Dicken 2016; and Vickers 2018, 49). It has been described as the strongest argument in favor of anti-realism, and the greatest challenge to realism. This problem arises from reflecting on the nature of theory change.

Kuhn's account of science, which insists that radical changes of theory are part of the normal developmental cycle of any scientific field, brought the issue into the spotlight. As Mary Hesse (1980) notes, "in the wake of Kuhn's *Structure of Scientific Revolutions*, many studies have laid emphasis on the revolutionary conceptual changes that take place in the sequence of theories in a given domain of phenomena" (Hesse 1980, x). And Hesse explicitly links this to Kuhn's work (also Hesse 2002; and Hallberg 2017, 166–167).[14] Indeed, so does John Worrall, who discusses the Pessimistic Induction under the telling label of "the argument from scientific revolutions" (see Worrall 1989, 99, but also 103). Worrall thus makes clear the connection to Kuhn. Addressing the Pessimistic Induction is now one of the central concerns of those engaged in the contemporary realism/anti-realism debates.

---

[14] Hesse adds that radical changes of theory cannot be reconciled with the view that "*accumulating data plus coherence conditions ultimately converge to true theory*" (Hesse 1980, viii and x). Elsewhere Hesse suggests that the Strong Programme's symmetry thesis also supports claims against privileging the theories we accept today (see Hesse 1982a, 328–329).

That Kuhn and the problem of theory change are responsible for the focus on the Pessimistic Induction, as I claim, explains the renewed attention on Henri Poincaré's writings in the contemporary realism/anti-realism debates. It is widely recognized that Poincaré presented a version of the Pessimistic Induction in the early 1900s, where he refers to the "ruins piled upon ruins" and "the bankruptcy of science" (see Poincaré 1903/2001, 122; also Worrall 1989; and Wray 2015a). The early Vienna Circle logical positivists read and were influenced by Poincaré (see Frank 1949/1961, § 4 and § 5). But Poincaré was largely ignored by those involved in the debates of the 1950s and 1960s, and his Pessimistic Induction was never discussed in that context. The rediscovery of Poincaré's Pessimistic Induction thus seems to be a consequence of the concern with the problem of theory change. And Worrall's subsequent casting of Poincaré as a Structural Realist is a consequence of realists attempting to address the threat posed by the Pessimistic Induction (see Worrall 1989).

As mentioned, Kuhn was not a major participant in the realism/anti-realism debates that he spawned. Nor did he contribute to an assessment of the Pessimistic Induction.[15] Instead, Kuhn got embroiled in a different set of debates concerning the *rationality* of science. As discussed in Chapter 9, a number of early critics argued that Kuhn's analysis of theory change raised serious questions about the rationality of science (see especially Scheffler 1967; Hall 1970; Lakatos 1970/1972; Laudan 1984; but also Friedman 2001, 47–50). The notion of incommensurability, specifically what has come to be called "meaning incommensurability," seems to be responsible for this development. In fact, some uncharitable readers of *Structure* claimed that Kuhn was saying that competing theories cannot even be compared, given that they do not refer to the same things, even when they use the same terms (see also Davidson 1973–1974). Consequently, evaluating their relative merits was impossible or futile. Hacking goes so far as to claim that Kuhn's book "produced a decisive transformation and unintentionally inspired a *crisis* of rationality" (Hacking 1983, 2; emphasis added).

Kuhn's problems in the rationality debates were amplified significantly when the Strong Programme in the Sociology of Scientific Knowledge (SSK) emerged. As we saw in Chapter 6, these sociologists identified as both Kuhnians and relativists (see, for example, Barnes 1982). In doing so,

---

[15] I have recently argued that Kuhn can offer some insight into the contemporary debate, insofar as he gives us a clear statement on what is involved in a radical change of theory, which is relevant to assessing the Pessimistic Induction (see Wray 2018a, Chapter 7).

they helped secure a firm association between Kuhn and relativism, which just further reassured Kuhn's critics that he was as dangerous as they had thought. The Strong Programme saw in normal science the same sorts of epistemic problems that Kuhn's philosophical critics saw in paradigm changes, or changes of theory. As we saw earlier, on the Strong Programme's reading of Kuhn, every single application of a concept is underdetermined by evidence and logic (see, for example, Barnes 1982, especially Chapter 2). But these sociologists of science were not distressed by this insight, unlike many philosophers who read *Structure*.

As I have argued earlier, as surprising as this may be, Kuhn did not anticipate this criticism, as he never doubted the rationality of science. As Kuhn reported in the 1980s, "the question that more than any other has guided and motivated me is . . . why the special nature of group practice in the sciences has been so *strikingly successful* in resolving the problems scientists choose" (Kuhn 1983a, 28; emphasis added). Thus, Kuhn took the success of science as given. It was his explanandum (or explicandum). But Kuhn ended up spending much of his career addressing the early critics who saw his view as a threat to the rationality of science.

Let me reiterate the nature of Kuhn's influence on the contemporary realism/anti-realism debates. Even though he was not concerned with the realism/anti-realism debates that were then going on when he wrote *Structure*, Kuhn's focus on theory change has left a significant mark on the contemporary realism/anti-realism debates. Kuhn made theory change an epistemic problem, a problem that realists had to contend with.

Kuhn's Remarks on Realism, Post-*Structure*

So far, I have argued that *Structure* played a crucial role in giving new direction to the contemporary realism/anti-realism debates. In the 1970s and 1980s, in a number of papers, Kuhn made some remarks that are at least prima facie relevant to the realism/anti-realism debates. It is worth examining these briefly. As we will see, often in these later papers Kuhn's concerns are tangential to the contemporary realism/anti-realism debates. What is more, even when his remarks are relevant to the contemporary debates, Kuhn is often concerned with other matters.

First, Kuhn has had a significant and undeniable influence on the debate about the theoretical values. His often-cited paper "Objectivity, Value Judgment and Theory Choice" provides an analysis of these values: simplicity, breadth of scope, consistency, accuracy and fruitfulness (see Kuhn 1973/1977). Kuhn describes these values as "*the* shared basis for theory choice" (Kuhn 1973/1977, 322; emphasis in original). And he suggests that scientists have appealed to such values throughout the history of science and across a range of scientific fields. This list of values has had a profound influence in the contemporary realism/anti-realism debates (see, for example, Schindler 2018). It has become a canonical list of sorts.

But Kuhn's purposes in discussing these values were somewhat tangential to the realism/anti-realism debates. Kuhn argues that these values, though widely accepted in the sciences, do not enable scientists to unequivocally resolve disputes about competing theories. As he notes, (i) each of the values is, itself, open to multiple interpretations, and (ii) collectively, the values can lead to divergent evaluations of competing theories (see Kuhn 1973/1977, 322). As a result, two scientists could each appeal to simplicity, for example, but could disagree regarding which of two different theories is the simpler, as each theory may be simpler than the other in different respects. And even if two scientists agree on which of two competing theories is simpler, broader in scope, more accurate, and so on, they could still disagree about which theory is superior, due to the fact that they weigh the values differently (see Kuhn 1973/1977, 331).

Ultimately, Kuhn's point in this discussion is that the malleability of the theoretical values is not a problem for science. In fact, he argues that it is an asset. That is, rather than being an impediment to the advancement of science, the malleability of these values plays an important role in ensuring that competing theories are developed. In Kuhn's words, "what from one viewpoint may *seem* the looseness and imperfection of choice criteria . . . [are] an indispensable means of spreading the risk which the introduction or support of novelty always entails" (Kuhn 1973/1977, 332; emphasis added). This is Kuhn's famous risk-spreading argument.

Strictly speaking, none of this speaks in favor of either realism or anti-realism. But realists have traditionally assumed that the theoretical values track the truth (see, for example, McMullin 1992). Consequently, Kuhn's analysis has been regarded as either supporting anti-realism or indicating Kuhn's own commitment to an anti-realist view (see especially McMullin 1992, 70–71). Kuhn's point, though, is that even if they did track the truth, the theoretical values cannot resolve disputes in a straightforward manner, as many assume.

In "Metaphor in Science," a paper written not long after "Objectivity, Value Judgment and Theory Choice," Kuhn claims to be a realist. He compares himself to Richard Boyd, and claims that "both [he and Boyd] are *unregenerate realists*" (Kuhn 1979/2000, 203; emphasis added). Boyd has been an important player in the realism/anti-realism debates, and he is without a doubt a realist. One of Boyd's most influential contributions to the debates is his argument that the role our background theories play in theory evaluation precludes scientists being unreliable in the sort of manner that anti-realists seem to assume they are (see, for example, Boyd 1985).

Kuhn's remark, above, that he is a realist like Boyd, is followed by this curious qualifier: "our [that is, Kuhn's and Boyd's] differences have to do with the commitments that adherence to a realist's position implies. But neither of us has yet developed an account of those commitments" (Kuhn 1979/2000, 203). Kuhn then proceeds to analyze their differences. Most importantly, Kuhn says that he rejects Boyd's metaphor "that scientific theories cut [or can cut] nature at its joints" (Kuhn 1979/2000, 205). Kuhn insists that "this way of talking is . . . only a rephrasing of the classical empiricists' position that successive theories provide successively closer approximations to nature" (Kuhn 1979/2000, 205). This is something that Kuhn had argued against from the start. He insisted, from the beginning, that the growth of scientific knowledge was not cumulative (Kuhn 1962/2012, 2). Kuhn then made an appeal to the history of science, arguing that he saw "no historical evidence for a process of zeroing in" (see Kuhn 1979/2000, 206). Clearly, he was rejecting Boyd's brand of realism.

Though Kuhn's remarks in this paper are clearly relevant to the realism/anti-realism debates, it is far from clear why he insists that he is a realist like Boyd. Boyd really does seem to get at a conviction that is central to realism, or at least to many forms of realism. That is, many realists do think science is zeroing in on the truth, to use Kuhn's expression. Kuhn's paper on metaphor, though, has not played an important role in the realism/anti-realism debates.

Kuhn also briefly discusses his relationship to realism in "Possible Worlds in the History of Science" (see Kuhn 1989/2000). Kuhn suggests that the development of science does not provide support for realism. Now, though, his argument is cast in terms of lexical changes, for he now characterizes revolutionary changes of theory as lexical changes that violate the no-overlap principle (see Kuhn 1989/2000, 76). The realist, Kuhn suggests, assumes that "the progress of science . . . [consists] in ever closer specification of a single world, the actual or real one" (Kuhn 1989/2000, 76). Instead, Kuhn argues that different scientific lexicons give access to a different set of possible worlds. Hence, as far as he is concerned, "scientific development turns out to depend not only on weeding out candidates for reality from the current set of possible worlds, but also upon occasional transitions to another set [of possible worlds], one made accessible by a lexicon with a different structure" (see Kuhn 1989/2000, 76). So, the fact that scientific development is periodically interrupted by significant lexical changes undermines realism. There is little reason, he claims, to believe that the development through such lexical changes constitutes a convergence on "the actual or real" world. Such changes, Kuhn explains, "are not well described by phrases like 'marginal adjustments' or 'zeroing in,'" contrary to what the realist would like us to believe (see Kuhn 1989/2000, 85).

Kuhn suggests that historians can more readily see this, as they experience the discontinuities across lexical lines when they attempt to understand scientific practices from earlier periods in which an alternative scientific lexicon was in use (see Kuhn 1989/2000, 76).

Here, Kuhn provides the clearest statement of what he takes realism to entail. He explains that "standard forms of realism presuppose ... [that] a statement's being true or false depends simply on whether or not it corresponds to the real world – independent of time, language, and culture" (see Kuhn 1989/2000, 77). But Kuhn insists that the "evaluation of a statement's truth values is ... an activity that can be conducted only with a lexicon already in place, and its outcome depends upon that lexicon" (see Kuhn 1989/2000, 77). Kuhn, though, leaves this debate unresolved, noting that it is "a task for another paper" (see Kuhn 1989/2000, 77). This was not a paper that Kuhn ever wrote.

Finally, Kuhn also makes some remarks in "The Road since *Structure*" that seem to have some bearing on his thoughts on realism, but here he contrasts realism with *constructivism* (see Kuhn 1991/2000a). After presenting his new account of scientific revolutions in terms of lexical changes, Kuhn ends the paper outlining "the relationship between the lexicon – the shared taxonomy of a speech community – and the world the members of that community jointly inhabit" (Kuhn 1991/2000a, 101). He insists that "it [that is, the relationship between the lexicon and the world] cannot be the one Putnam has called metaphysical realism" (Kuhn 1991/2000a, 101). But when Kuhn tries to articulate exactly what his view entails things become somewhat elusive and muddled. He does say "the world is not invented or constructed," for we "find the world already in place" (Kuhn 1991/2000a, 101). Further, he notes that "it is entirely solid: capable of providing decisive evidence against invented hypotheses which fail to match its behaviour" (Kuhn 1991/2000a, 101). But then he notes that, "insofar as the structure of the world can be experienced ... it is constrained by the structure of the lexicon of the community that inhabits it" (see Kuhn 1991/2000a, 101).

Kuhn insists that the view he is presenting here is consistent with a robust form of realism. He asks, rhetorically, "can a world that alters with time and from one community to the next correspond to what is generally referred to as 'the real world'?" (Kuhn 1991/2000a, 102) And he makes clear that he thinks it can be. "It provides the environment, the stage, for all individual and social life. On such life it places rigid constraints ... What more can reasonably be asked of a real world?" (Kuhn 1991/2000a, 102). Kuhn ends this discussion characterizing his view as "a sort of post-Darwinian Kantianism" (see Kuhn 1991/2000a, 104). Unpacking this label, Kuhn explains that, "like Kantian categories, the lexicon supplies preconditions of possible experience. But lexical categories, unlike their Kantian forebears, can and do change" (Kuhn 1991/2000a, 104). Finally, he notes that "underlying all these processes of differentiation and change, there must ... be something permanent, fixed and stable. But, like Kant's *Ding an sich*, it is ineffable, undescribable, undiscussable" (Kuhn 1991/2000a, 104).

Much of what Kuhn said about realism after the publication of *Structure* is largely tangential to the concerns that divide those involved in the contemporary realism/anti-realism debates. Perhaps one of the reasons that Kuhn did not engage directly in or influence the contemporary debates much is because his claims about realism are not well articulated.[16] In fact, some of his remarks even seem incoherent. But this does not diminish the impact that the problem of theory change has had on the contemporary debates. And this is an important part of Kuhn's philosophical legacy.

Situating Kuhn in the Contemporary Debates

Even though Kuhn did not actively engage in the realism/anti-realism debates, this has not stopped others from either discussing the relevance of his views to the debates or attempting to situate him in the current debates. It is worth briefly examining the variety of ways in which Kuhn's views are discussed in these contexts. Though somewhat selective, my analysis will offer additional support for my argument that Kuhn

---

[16] As one referee for Cambridge University Press noted, Kuhn also discusses realism and related topics in his Thalheimer Lectures, delivered in 1984. The topic is discussed in Lecture IV, "Conveying the Past to the Present" (see Kuhn 1984b). But here Kuhn slides between (i) drawing attention to aspects of his view that align with a realist position and (ii) suggesting that his view aligns with anti-realism. He even acknowledges that he understands "better why [his] name occurs so often in Rorty's pages" (Kuhn 1984b, 119).

contributed to defining the problem-space that shapes the contemporary realism/anti-realism debates.

In the early 1990s, as a contribution to a Festschrift honoring Kuhn, Ernan McMullin argued that "Kuhn's influence on the burgeoning anti-realism of the last two decades can scarcely be overestimated" (see McMullin 1992, 71). McMullin argues that Kuhn has been widely misunderstood as attacking the rationality of science. Instead, McMullin insists that Kuhn's arguments are really aimed at attacking the alleged "truth character of theories" (see McMullin 1992, 70). As McMullin explains, "Kuhn ... rejects in a most emphatic way the traditional realist view that the explanatory success of a theory gives reason to believe that entities like those postulated by the theory exist, i.e., that the theory is at least approximately true" (see McMullin 1992, 70). McMullin thus regards Kuhn as an anti-realist.

Theodore Arabatzis also argues that Kuhn clearly comes out on the anti-realist side of the debates. Arabatzis suggests that Kuhn's focus on the incommensurability between successive theories in a field supports anti-realism (see Arabatzis 2001, § 4). He claims that in his later writings, Kuhn "made it clear that incommensurability is incompatible with a realist position" (Arabatzis 2001, S538). Further, Arabatzis argues that "aspects of the historical development of science ... undermine certain realist arguments and support an anti-realist outlook" (Arabatzis 2001, S540). Arabatzis has in mind what I call the problem of theory change.

For similar reasons, John Worrall also suggests that Kuhn "explicitly rejects any form of scientific realism" (Worrall 2002, 91). Worrall claims that Kuhn "found it impossible to see in actual cases of successive theory changes from the history of science anything like a consistent movement towards greater 'approximate truth'" (Worrall 2002, 91). Worrall, though, also rightly notes that Kuhn did not think that this undermined or threatened the claim that science progresses (see Worrall 2002, 91). As Worrall explains, Kuhn believed that theories get better and better with each successive change of theory. But he denies that "'better and better' here means 'truer and truer'" (Worrall 2002, 91).

Gerald Doppelt also regards Kuhn as an anti-realist. He describes Kuhn's view as "'a paradigm account' of the success of science that makes *no* realist assumptions" (Doppelt 2013, 44; emphasis added). Doppelt also refers to "widely accepted features of theory-change ... brought onto center stage by Kuhn's work" (Doppelt 2013, 43). Specifically, Doppelt claims the *central problem for realists* is that many theories in the history of science are successful in their time [e.g. the phlogiston theory of combustion, caloric theory of heat, the ether theory of light, etc.] even though their key ontological entities and claims about unobservables are rejected or "falsified" by the theories in a field which supersede them. (Doppelt 2013, 43; emphasis added)

Indeed, Doppelt seems astutely aware of the wide-ranging impact of these observations about theory change, and their relevance to the contemporary realism/anti-realism debates. He claims

this fact of ontological and conceptual discontinuity between successful theories in a field motivates [i] the anti-realist arguments of Laudan (the pessimistic meta-induction), [ii] the revisionary but still "ontological realism" of Psillos, and [iii] the structural realism of Worrall, Ladyman, Carrier, and others. (Doppelt 2013, 43; numerals added)

Ultimately, Doppelt appeals to Kuhn's paradigm account as a foil in his presentation of his own Best Current Theory Realism (see Doppelt 2013, § 5). Thus, Kuhn occasionally figures in the contemporary debates as a useful foil against which to define one's own position in the debate, much as instrumentalism functioned in the debates of the 1950s and 1960s.

Not all philosophers who discuss Kuhn in the context of the contemporary realism/anti-realism debates characterize him as an anti-realist. Ron Giere, for example, claims that Kuhn is a realist, specifically, a perspectival realist, at least in his later writings, published in *The Road since Structure* (see Giere 2013, 53). But Giere also acknowledges that Kuhn denies that successive theories in a field are getting ever closer to the truth (see Giere 2013, 55). Giere's reading of Kuhn, however, has not gone uncontested. Michela Massimi, for example, is less sanguine about the prospects of casting Kuhn as a realist as Giere has done (see Massimi 2015). Specifically, she notes that there are significant challenges to reconciling Kuhn's talk of world changes with realism.[17]

Howard Sankey provides the most thorough and systematic analysis of Kuhn's relationship to realism. Sankey argues that Kuhn's view is more in line with an anti-realist view. But Sankey also identifies a number of claims often taken to be incompatible with realism that are associated with Kuhn but are in fact compatible with realism. These include: (i) "that the

---

[17] Kuhn's discussion of "revolutions as changes of world view" seems relevant to the realism/anti-realism debate, but it has been read in the most uncharitable ways. Peter Godfrey-Smith, for example, claims that "the X-rated Chapter X [of *Structure*] is the worst material in Kuhn's great book. It would have been better if he had left this chapter in a taxi" (Godfrey-Smith 2003, 96).

evaluation of scientific theories is multi-criterial" (Sankey 2018b, 82); (ii) "that scientific progress may be thought of in evolutionary and problem-solving terms" (82); and (iii) "that substantial conceptual change takes place in the history of science" (82).

Finally, the historian of science Peter Gordon argues that Kuhn distanced himself from anti-realism, especially in his later publications. As Gordon explains, "Kuhn himself did not intend, and later came to regret, some of the more extravagant versions of anti-realism or radical constructivism ... ascribed to his book" (Gordon 2012, 128). Indeed, Gordon notes that Kuhn's provocative remarks about paradigm changes involving world changes are always qualified in such a manner as to reduce their anti-realist implications (see Gordon 2012, 128–130).

My purpose here is not to provide some sort of final judgment in the debate about whether to classify Kuhn as a realist or an anti-realist. Rather, my point in reviewing this literature has been to show how profoundly Kuhn's problem of theory change set the terms of the contemporary realism/anti-realism debates.

In summary, as we have seen, the current realism/anti-realism debates are quite different from the debates of the 1950s and 1960s. In the earlier debates, anti-realists were identified with an instrumentalist view of scientific theories, and those who were identified as espousing such a view were physicists, not philosophers. A significant shift occurred around 1970 or so, in the wake of the publication of Kuhn's *Structure*. As Paul Dicken claims,

> following the demise of logical empiricism, the contemporary scientific realism debate evolved from an essentially semantic issue concerning the language of our theories to an epistemological issue concerning the extent to which we are justified in believing those theories to be true. (Dicken 2016, 91)[18]

I have argued that Kuhn's *Structure*, with its focus on radical theory change, is responsible for this shift. Kuhn made theory change a problem, and one that realists felt compelled to address. The fact, or alleged fact, that successive theories in a field have radically different

---

[18] Matthias Neuber claims that Herbert Feigl was instrumental in setting the terms of the earlier realism/anti-realism debates, the ones that were primarily concerned with semantics, not epistemology (see Neuber 2011, 171). Apparently Feigl referred to his own view as "semantic realism" (see Neuber 2011, 172). Like Smart, Maxwell and Popper, Feigl believed that "theoretical terms were supposed to refer to unobservable, mind-independent entities, so that, for example, the referent of the term 'atom' would be real atoms and not samples of 'logical constructions' out of sense data (or other kinds of directly perceivable things)" (see Neuber 2011, 173).

ontologies seems to pose serious challenges to many forms of realism. This is a significant part of his philosophical legacy. And even though Kuhn was not a central player in such discussions, in part because he did not articulate a consistent position on realism or anti-realism, many have attempted to situate his view in these debates.

CHAPTER 11

*Kuhn's Career, Post-*Structure

Throughout this book, I have been examining (i) the various influences on Thomas Kuhn as he wrote *Structure*, as well as (ii) the subsequent impact of the book across a range of disciplines. In closing, I want to survey what Kuhn did after *Structure*, and the relationship these projects had to *Structure*.

First, I want to scrutinize some remarks Kuhn made on both (i) how he might have improved *Structure* were he to revise it, and (ii) what he conceived a longer, though ultimately unwritten version of *Structure* would have contained. Second, I examine the various projects he engaged in after the publication of *Structure*. Some of these have already been discussed in some detail in previous chapters. Finally, I will briefly remark on Kuhn's unfinished book manuscript. Kuhn was working on a book when he died, and though plans to have it published were made before his death, it has still not been published.

Structure Revised and Republished

Structure was originally published in 1962 both as a contribution to the *Encyclopedia of Unified Science* and as a stand-alone monograph (see Kuhn 1997/2000, 300–301). It was reprinted in 1970, with the addition of a Postscript, and then again in 1996. The third edition was the first time an index was included. Kuhn once remarked: "Challenged to explain the absence of an index, I regularly pointed out that its most frequently consulted entry would be: 'paradigm, 1–172, passim'" (Kuhn 1970/1977, 294). In fact, around the time the third edition of *Structure* was being prepared for publication, someone sent to Kuhn an unsolicited index they had constructed (see Abrams 1995a). This unsolicited index formed the basis of the index that was published in the third edition.

A fourth edition was published in 2012, on the fiftieth anniversary of the book, which includes an introductory essay by Ian Hacking, one of Kuhn's most sympathetic readers. Probably rightly so, Kuhn never attempted to rewrite *Structure*. In the Postscript to the second edition, though, Kuhn makes some remarks that suggest that he was considering revising the book. In a footnote, for example, he makes the remark that "for this edition I have attempted no systematic *rewriting*" (see Kuhn 1969/2012, 173, fn. 2; emphasis added).[1]

In the Preface to *Structure*, Kuhn suggests that an expanded version of the book may be on the horizon. He explains that the book's brevity is a consequence of its being published as a contribution to the *Encyclopedia of Unified Science*. The various contributions to the *Encyclopedia* were generally quite short, and certainly shorter than *Structure*. Thus, Kuhn notes that "this work [that is, *Structure*] remains an essay rather than the full-scale book my subject will ultimately demand" (Kuhn 1962/2012, xliii). The next sentence seems to be more explicit in alluding to a longer book in the works. Specifically, Kuhn remarks that, "since my most fundamental objective is to urge a change in the perception and evaluation of familiar data, the schematic character of *this first presentation* need be no drawback" (Kuhn 1962/2012, xliii; emphasis added). Indeed, that he intended to write a longer version of the book is supported by remarks he made in a letter to the editor he was working with at the University of Chicago Press. Noting that his manuscript is much longer than the other contributions to the *Encyclopedia of Unified Science*, Kuhn remarks that

> the manuscript now in your hands contains most of what I've got to say about my subject at this time. I still expect to do [a full-length book], but only after collecting reactions to this version and also completing a good deal more research and cogitation. Probably the "definitive" version is a decade or more in the future. (Kuhn 1961a)

Clearly, Kuhn had conceived of there being a second, larger, presentation of these ideas.

In the Preface to *Structure*, Kuhn identifies "the sorts of extension in both scope and depth that [he hoped] ultimately to include in a longer version" (Kuhn 1962/2012, xliii). First, he suggests that there is more, and more varied, historical evidence he could present, including evidence from the biological sciences, which are rarely discussed in *Structure* (see Kuhn 1962/2012, xliii).[2] He also notes that the brevity of the book made it

---

[1] In the Postscript written for the second edition, Kuhn notes that he is "restricting alterations to a few typographical errors plus two passages which contained isolable errors" (Kuhn 1969/2012, 173, fn. 2).

[2] Indeed, a number of biologists and historians and philosophers of biology have taken it upon themselves to assess the extent to which Kuhn's account of science fits their own discipline. See,

necessary to "forego the discussion of a number of major problems" (Kuhn 1962/2012, xliii). These problems are noteworthy because they make clear that Kuhn was quite aware of how his presentation made the whole cycle of scientific change look a bit too clean and straightforward.

> My distinction between the pre- and the post-paradigm periods in the development of a science is ... much too schematic. Each of the schools whose competition characterizes the earlier period is guided by something much like a paradigm; there are circumstances, though I think them rare, under which two paradigms can coexist peacefully in the later period. Mere possession of a paradigm is not quite a sufficient criterion for the developmental transition discussed in Section II [of *Structure*]. (Kuhn 1962/2012, xliii–xliv)

With these qualifying remarks Kuhn makes it clear that he was anticipating some common misunderstandings. Many critics have taken issue with Kuhn's account because they are able to identify examples from the history of science where (i) it is not so easy to determine whether a field is still in a pre-paradigm state, or (ii) two paradigms exist in a field that is allegedly past the pre-paradigm stage, or (iii) a field has settled on a paradigm but is not working through the cycle that Kuhn suggests characterizes fields once they have left the pre-paradigm stage of development. These examples are taken to show that Kuhn's theory is mistaken. But, as far as Kuhn is concerned, none of these conditions are precluded by his account of scientific change. He just believes that these sorts of cases are atypical in the natural sciences.

Kuhn also notes that "except in occasional brief asides, [he has] said nothing about the role of technological advance or of external social, economic, and intellectual conditions in the development of science" (Kuhn 1962/2012, xliv). These sorts of issues, he suggests, would be addressed in a longer version of the book. Here, though, Kuhn notes

---

for example, Ernst Mayr (1994), Adam Wilkins (1996) and Vincenzo Politi (2018). As Mayr makes clear, he believes that "scientific revolutions in biology do not conform to the description of such revolutions as given by T. S. Kuhn" (Mayr 1994, 328). Contrary to Kuhn's account, in biology "several paradigms may coexist simultaneously for long periods of time, and succeeding paradigms are not necessarily incommensurable" (Mayr 1994, 328). Further, "biological revolutions are not separated by long periods of normal science" (Mayr 1994, 328). Wilkins claims that "the Darwinian and the molecular biological revolutions do not comfortably fit the Kuhnian model of a scientific revolution" (Wilkins 1996, 696). He also suggests that the Mendelian revolution does not fit the mold either (see Wilkins 1996, 696). Wilkins focuses a lot on Kuhn's claims about the age of scientists in accepting new theories in his evaluation of these revolutions in biology. But Kuhn's empirical claims about (i) age and discovery and (ii) age and the acceptance of new theories have been shown to be false (see Wray 2003). Recently, Petter Portin (2015) has examined the development of the field of genetics through a Kuhnian lens (see Portin 2015).

that "explicit consideration of effects like these would not . . . modify the main theses developed in this essay, but it would surely add an analytic dimension of first-rate importance for the understanding of scientific advance" (Kuhn 1962/2012, xliv). Thus, he recognizes that scientific research communities are not wholly shielded from the influence of the broader cultures. What Kuhn took to be important is that such external factors do not determine how competing theories are evaluated in the natural sciences. Rather, the experts in a field determine how competing theories are assessed, for only they are in a position to adequately assess competing theories, and evaluate the contributions of their peers.

Kuhn also notes that, "perhaps most important of all, limitations of space have drastically affected my treatment of the philosophical implications of this essay's historically oriented view of science" (Kuhn 1962/2012, xliv). Exactly which philosophical implications he had in mind, was left unsaid, as the longer version of *Structure* that Kuhn alludes to in the Preface was never written.[3]

Interestingly, both in his replies to his critics at the Bedford College conference and in the Postscript to the second edition of *Structure*, Kuhn makes some passing remarks about how he *might* have written *Structure* differently. Here, the implication is that he is no longer planning to write a longer version of the book. His remarks in the Postscript were made after the book had endured seven years of critical scrutiny, so he had some sense of where people had misunderstood him. For example, Kuhn notes that "a new version of *Structure* would open with a discussion of community structure" (Kuhn 1970/2000, 168; see also Kuhn 1969/2012, 175). He acknowledges the work that sociologists were then doing on this issue, including the then-recent work on invisible colleges (see Kuhn 1969/2012, 175, fn. 5). He also suggests that "I now think it is a weakness of my original text that so little attention is given to such values as internal and external consistency in considering sources of crisis and factors in theory choice" (Kuhn 1969/2012, 184).

But, given the book's success, it is not surprising that he remained quite satisfied with the book. In fact, near the end of his life, Kuhn remarked

> clearly, I wanted it to be an important book; clearly it was being an important book . . . but on the other hand, I recognized that if I had to

---

[3] In an interview conducted near the end of his life, Kuhn mentions that he had conceived of the book he agreed to write for the *Encyclopedia of Unified Science* as "the first version, a short version of *The Structure of Scientific Revolutions*" (Kuhn 1997/2000, 292).

> do it all over again, I could, if I had the opportunity, eliminate some of the misunderstandings. But if I couldn't do that, I'd do it all over again the way it was. I mean, I had disappointments, I didn't have regrets. (Kuhn 1997/2000, 309)

It is questionable whether he could have done better than he had done with *Structure*. As we have seen, *Structure* has been a remarkably influential book. It has been a stimulating source of many fruitful ideas across a wide range of disciplines. And it has given us, both academics and educated laypeople, a vocabulary for talking about ideas and intellectual changes.

Susan Abrams, an editor Kuhn had worked with from the University of Chicago Press, gives a clear sense of the magnitude of the audience of *Structure*, even thirty-three years after its publication. In a letter from 1995, Abrams remarks that "something akin to panic strikes when we [that is, the University of Chicago Press] have fewer than 6,000 copies in stock" (Abrams 1995a). Few academic philosophy books will ever sell 6,000 copies, let alone have the need for a publisher to have 6,000 copies in stock at all times! In another letter from 1995, Abrams remarks that "we continue to sell between 22,000 and 26,000 copies a year" (Abrams 1995b). Again, that an academic book can sell this well is astounding.

Not surprisingly, Kuhn's sense of the importance of the book, as well as his self-conception as the author of the book, changed throughout the years. But it is interesting to consider how he felt about matters in 1966, just four few years after the publication of *Structure*. At that time, he was teaching at Princeton. Kuhn provides some interesting insight in his lecture notes. He began his course with some autobiographical remarks in which he discussed the challenges of classifying the book. As these remarks are from his lecture notes, they are not always in full sentences. The lecture notes that are relevant to the book read as follows:

3. As most or all of you know I published about three years ago a book which has since been raising some dust in a number of quite different circles.

   a. When I wrote [it] I was not quite certain what field it was in and nothing that has happened since has much clarified [the] problem.
      i Am still at a total loss if someone asks at a cocktail party "What is your book about?" . . .
   b. I'm an ex-physicist, turned professionally to hist. [of] science.
      i Book reflects experience in both fields, but is certainly in neither.

c. Equally reflects longstanding interest in philosophy of science, one that lasted until I took to history.

    i. But in no usual sense is it a book in [the] phil. of sci.

    . . .

    iv Not written in phil. language or with real background in the literature. References or vaguely to a tradition and surely not really fair.

d. Probably, then, best description of the book is sociology of science.

    i. But not clear that that field quite exists yet or what it will be like if it does. . . .

4. This ambiguity has left me with a problem ever since [the] book was published.

    . . .

    e. As I've gone on [I] have increasingly realized that what I myself most want to get at is the philosophy in the book. (Kuhn 1966b)

Clearly, he was delighted that the book had raised some dust, no matter what sort of book it was.

After *Structure*

It is hard to follow up on a book as successful as *Structure*. But it is worth tracing the various projects that Kuhn embarked on after its publication. Some of these were great successes, but many have failed to appreciate this because we compare them to *Structure*, which was successful beyond all expectations.[4]

Immediately following the publication of *Structure*, and just before he embarked on the History of Quantum Physics Project in Copenhagen, Kuhn was thinking of writing a "book about science and philosophy in the seventeenth century" (Kuhn 1997/2000, 303). Specifically, he wanted to write a book about the separation of philosophy from science in the early modern period (see Kuhn 1997/2000, 290). The opportunity to work on the archival project pushed this book idea to the side. And, as Kuhn explains, "I got too busy with other things – that book is never going to appear from my hand" (Kuhn 1997/2000, 290).

---

[4] Stanley Cavell remarked that when Kuhn was still working on *Structure*, Cavell "firmly believed [*Structure*] would make [Kuhn] famous. (Not *that* famous)" (Cavell 2010, 356–357; emphasis in original).

But since the publication of the second edition of *Structure*, there have been two edited volumes of Kuhn's essays published. The first, *The Essential Tension*, was published in 1977. It includes a mix of papers on (i) history and historiography of science, and (ii) the philosophy of science. Then in 2000, four years after Kuhn's death, a collection of papers was published under the editorship of James Conant, who was James B. Conant's grandson, and John Haugeland. This volume, *The Road since Structure*, contains mostly philosophical papers, but it also includes a rare and insightful interview that Kuhn gave near the end of his life. Kuhn apparently decided which papers were to be included in the volume, and which were not to be included (see Conant and Haugeland 2000, 1).

Shortly after the publication of the first volume of papers, *Essential Tension*, Kuhn published his final monograph in the history of science, the book on Planck, *Black-Body Theory and the Quantum Discontinuity*, discussed in detail earlier in Chapter 8.

It is also worth briefly mentioning another project Kuhn alludes to on a number of occasions. This is a computer program he was working on, one that models the grouping of objects according to their similarities. It seems that Kuhn must have been working on this project from the late 1960s into the mid-1970s. In the Postscript to *Structure* Kuhn briefly discusses the computer program. He mentions it in the context of addressing a particular criticism raised against *Structure*, specifically, that he "was trying to make science rest on unanalyzable individual intuitions rather than on logic and law" (Kuhn 1969/2012, 190). This criticism, Kuhn suggests, was connected with the broader "charges of subjectivity and irrationality" (Kuhn 1969/2012, 190).

Kuhn believed that the criticism was baseless, and he explained that he was "currently experimenting with a computer program designed to investigate [the] properties" of the intuitions that underlie scientists' classification of phenomena (see Kuhn 1969/2012, 191). The computer program was intended to model the process by which a scientist acquires "from exemplars the ability to recognize a given situation as like some and unlike others that one has seen before" (Kuhn 1969/2012, 191). What he was aiming to investigate was the possibility that scientists could group things without relying on knowledge of necessary and sufficient conditions of membership in the classes or kinds. Indeed, this is one of the roles that he felt that paradigms play in science. That is, paradigms are a substitute of sorts for knowledge of the necessary and sufficient conditions for membership in the classes to which scientists assign things when they are classifying them. Kuhn's concern was that kinds defined in terms of necessary and sufficient conditions would be too rigid, and would not admit of the sort of flexibility that one sees in scientists' reasoning when they are solving research problems. The sort of puzzle solving that characterizes normal scientific research, Kuhn notes, is not a mechanical or algorithmic process. The concepts used by scientists in solving these puzzles need to be malleable, for the sorts of phenomena to which they are applied do not precisely match the sorts of phenomena to which they have been applied in the past.

Kuhn refers the interested reader to his paper "Second Thoughts on Paradigms," which seems to have been written roughly concurrently with the Postscript to the second edition of *Structure* (see Kuhn 1969/2012, 191, fn. 12; and 196, fn. 14).[5] Here Kuhn discusses his computer program in relationship to a more prosaic example, the example of a young boy, Johnny, who is learning to classify birds, with guidance from his father. Though Johnny learns to distinguish swans from ducks, and both from geese, he does not learn the necessary and sufficient conditions for membership in the three classes: swans, ducks and geese. Instead, he becomes aware of criteria that enable *him* to discern one type from the other, criteria that may in fact differ from the criteria his father or anyone else uses to group the various birds encountered into the various classes, even though his father and others may group the same things in the same classes. Kuhn's point is that though two or more people may successfully group the same individuals in the same classes, it does not follow that they define the classes or kinds in the same way. The computer program Kuhn sought to develop was supposed to be able to mimic Johnny in this endeavor. That is, it would be given a set of clusters, corresponding to the classes of ducks, swans and geese that Johnny encountered, and some exemplars of each kind. But then, like Johnny, it was supposed to be able to "go on," that is, continue to classify hitherto-unseen birds into the appropriate classes (see Kuhn 1974/1977, 309–312).

Kuhn's aim, he explains, was to show that "shared examples have essential cognitive functions prior to a specification of criteria with respect to which they are exemplary" (Kuhn 1974/1977, 313). Kuhn wanted us to recognize "the integrity of a cognitive process like the one just [described]" (Kuhn 1974/1977, 313). That is, he wanted us to recognize that some process like this is indispensable to the scientist, and that scientists do not in fact

---

[5] In a footnote Kuhn notes that "the possibility of immediate recognition of the members of natural families depends upon the existence ... of empty perceptual space between the families to be discriminated" (Kuhn 1969/2012, 196, fn. 14).

need to have knowledge of the necessary and sufficient conditions of class membership for the kind terms they use.

Kuhn mentions the computer program again in his replies to his critics at the Bedford College conference, which were also written contemporaneously with these other two papers (see Kuhn 1970/2000, 171–172). Indeed, the idea for developing such a program may have originated at the conference, for there is an exchange of letters between Margaret Masterman and Kuhn from 1966, in which they briefly discuss the extent to which computers could model analogical reasoning (see Masterman 1966; and Kuhn 1966a).[6] But it seems that the program was not a success. Sometime shortly after 1974, Kuhn gave up on the project, recognizing his own limitations with respect to computer science. In a letter responding to an inquiry about the project, Kuhn reports:

> I learned a great deal from the attempt, and what I learned has continued to play a role in my thought. But I discovered fairly early that I would need to become an expert programmer and to have hours of computer time at my disposal if my experiment were to be carried much further. Shortly after [1974] . . . I dropped the project, and it had no published product. (Kuhn 1986)

When Kuhn talks about the challenges of programming and the computer time required to advance the project, it is worth remembering he is talking about the mid-1970s.

Though Kuhn abandoned the project, his attempt to model scientific reasoning with a computer program shows that his defense of his accounts of scientific change and scientific knowledge were not built wholly on evidence drawn from the history of science. Indeed, it is quite misleading to suggest that Kuhn thought the history of science could provide sufficient data from which to build a philosophy of science. In his appeals to the history of science, he was not an inductivist, merely surveying a sample of scientific revolutions, for example, and drawing conclusions about what all revolutions have in common.[7] Indeed, his attempt to develop this computer program shows another persistent pattern in Kuhn's theorizing about science. It is continuous with his engagement with the research literature in psychology and the cognitive sciences, a theme appreciated

---

[6] In the exchange, Kuhn objects to Masterman's suggestion that "a paradigm is somehow the same as a model and its relata is that of analogy" (Kuhn 1966a). Instead, Kuhn insists that "the paradigm is a concrete example, that other concrete examples of the same kind are also paradigms, and that the relation between them is similarity, taken as a primitive" (Kuhn 1966a).

[7] Adam Wilkins, for example, claims that "in contrast to many earlier, *a priori*, philosophical theories of knowledge, Kuhn built his case from examples, in effect inductively" (Wilkins 1996, 695).

by David Kaiser (see Kaiser 2016). Recall that, in *Structure*, Kuhn drew on the work of Jean Piaget, Bruner and Postman and other psychologists.[8] Indeed, in her 1963 review of *Structure*, Mary Hesse had noted Kuhn's "illuminating comparisons with the psychology of perception, [and] his use of evidence from the sociology of scientific groups" (see Hesse 1963, 287). In this respect, Kuhn was truly trying to develop a naturalized epistemology of science, even before the term "naturalized epistemology" was coined. And his interest in and engagement with research in psychology continued until the end of his life.

The Unpublished Manuscript

Kuhn scholars have long been aware that Kuhn was working on another monograph when he died. Indeed, this book has been alluded to for many years. As noted earlier, Kuhn was notoriously bad at estimating how long it would take him to complete projects. Indeed, even as an undergraduate, when he was working on *The Harvard Crimson*, the student paper, he was "finding it very hard to write" (see Kuhn 1997/2000, 264). Similarly, when he was working on the Lowell Lectures, Kuhn reports: "I had a dreadful time preparing [the lectures] and I nearly cracked up" (Kuhn 1997/2000, 289). During his Guggenheim Fellowship in the mid-1950s he was able to finish neither *The Copernican Revolution* nor "the monograph for the encyclopedia" (Kuhn 1997/2000, 292).[9] In an NSF grant application requesting support for eight months, from the beginning of January 1990 to the end of August 1990, Kuhn states that "during 1989–90 the principle investigator hopes to put into final form much of the book to which his research during the last ten years has been devoted. The book's present working title is *Words and Worlds: An Evolutionary View of Scientific Development*" (see Kuhn 1989). And near the end of his life Kuhn remarked: "I've never been any good in saying how long it will take me to do things, for ten years I have been saying, I think it will take me

---

[8] In fact, Kuhn was already reading Piaget's work in 1949. His notebook from that time indicates that he read both *Judgment and Reasoning in the Child* and *Les notions de mouvement et de vitesse chez l'enfant* (see Kuhn 1949).

[9] Kuhn notes that when he was teaching the History of Science courses at Harvard with Leonard Nash, after Conant gave up the courses, teaching also became challenging for him. He explains: "teaching began to be difficult for me . . . Now I started spending much too much time preparing, getting very nervous in advance, and I've never altogether gotten over that . . . which has cost me some things . . . probably including some facility and sitting down and writing easily, although writing was always different from talking for me" (Kuhn 1997/2000, 284).

about another two years to finish this book I'm still working on, and I still think it will take me another two years" (Kuhn 1997/2000, 292).

Consequently, given his struggles with writing, it is not a surprise that Kuhn did not manage to finish his final book. But, aware that he was dying, he arranged for James Conant and John Haugeland to bring the project to completion in some sense, and see that it was published. However, both his editor at the University of Chicago Press, Susan Abrams, and John Haugeland have since died, and at the time of writing (2020), the manuscript has still not been published.

In 1984, when Paul Hoyningen-Huene went to work with Kuhn at MIT in an effort to complete his own book on Kuhn's theory of science, Kuhn suggested that "[he, that is, Kuhn himself,] hoped ... that he [Kuhn] could finish his new book at least by the late 1980s" (see Hoyningen-Huene 2015, 190). Hoyningen-Huene discussed parts of Kuhn's manuscript with Kuhn at that time (see Hoyningen-Huene 2015, 191). Hoyningen-Huene also provides a sense of the content of Kuhn's projected book, by providing a Table of Contents. This includes a list of unfinished or unwritten chapters as well as those that were in good draft form. He also identifies specific papers that Kuhn had presented over the years and that figured in Kuhn's manuscript, in one form or another.

The projected book changed its title over the years: In 1984 it was "Scientific Development and Lexical Change"; by 1990 it was "Words and Worlds: An Evolutionary View of Scientific Development"; and by the end of Kuhn's life the working title was "The Plurality of Worlds: An Evolutionary Theory of Scientific Development" (see Hoyningen-Huene 2015, 190).[10] Indeed, it seems that by at least May 1993, Kuhn had settled on the title of "Plurality of Worlds" (see Kuhn 1993a).

In the new book, the notion of a lexicon figures significantly. Indeed, scientific theories are understood as lexicons. Phases of normal science are those periods in which the lexicon in a scientific field is stable, and scientific revolutions involve the replacement of one lexicon with a new incompatible, or incommensurable, lexicon (see Hoyningen-Huene 2015, 192). Kuhn was intending to criticize the causal theory of reference, developed by Saul Kripke and Hilary Putnam. He wanted to show that the theory "doesn't work across periods of revolution ... but it works very well between those [periods]" (Kuhn 1997/2000, 313). Despite his critical

---

[10] It is interesting to note that David Lewis, Kuhn's former colleague at Princeton, published a book with the title *On the Plurality of Worlds* in 1984. And, much earlier (in 1686), Bernard le Bovier de Fontenelle published a book titled *Conversations on the Plurality of Worlds*, which explains the Copernican cosmos to a wider audience.

appraisal of the causal theory, Kuhn acknowledges that it led him to "go back and think about the Copernican revolution – what you can't trace through it is 'planets.' Planets are just a different collection before and afterward. There was a sort of localized break" (Kuhn 1997/2000, 312). Indeed, Kuhn would appeal to the Copernican Revolution in later papers, emphasizing that both the extension and the intension of the term "planet" changed during the revolution (see, for example, Kuhn 1987/2000). He would also emphasize the local nature of the changes wrought by a scientific revolution.[11]

Significantly, Kuhn draws heavily on research in developmental psychology (see Hoyningen-Huene 2015, 193). Further, in a letter written in 1992, Kuhn remarks: "the book on which I'm at work will draw significant parallels between revolutions and the biological process of speciation" (see Kuhn 1992). Before this sentence, he notes, "I don't think of evolutionary development as opposed to revolutionary. On the contrary, I conceive scientific development as fully evolutionary and revolutions as one of the sorts of change that occur within it" (Kuhn 1992). Indeed, this is an important theme, one that is developed in a number of papers Kuhn wrote around this time (see, especially, Kuhn 1991/2000a and 1992/2000). I had not seen this letter when I wrote my earlier book, *Kuhn's Evolutionary Social Epistemology* (see Wray 2011). But it provides a vindication of a central claim in that book, that Kuhn was committed to the view that scientific revolutions are not incompatible with evolutionary development. Indeed, Kuhn underscored the importance of the evolutionary analogy in his later work in other letters he wrote at around the same time (see, for example, Kuhn 1991).

Though the public has no access to the manuscript, the evaluation of Kuhn's NSF application from 1989 suggests that it was a promising project. The Panel Summary reads as follows:

> The PI [Principal Investigator], as all reviews assert, is clearly the most influential historian and philosopher of science of the past third of a century. It is important for history, philosophy and sociology of science for the PI to take up again the issues raised in the *Structure of Scientific Revolutions* nearly 30 years ago. (NSF 1989)

Clearly, the social capital that Kuhn had acquired from writing *Structure* was still holding up well, even among his peers.

---

[11] Mary Hesse (1982b) also suggests that extensionalist theories of reference, like Putnam's causal theory, are problematic. Such theories fail to account for the fact that metaphorical uses of language, the sorts of things that Kuhn was getting at in his discussions of scientists applying paradigms, undermine the power to deduce consequences from scientific theories.

Bibliography

Abbott, A. 2001. *Chaos of Disciplines*. Chicago, IL: University of Chicago Press.

Abrams, S. 1995a. "Letter to Professor Thomas S. Kuhn – Dated 13 March 1995," in the TSK Archives. Box 21: Folder 1, Correspondence, Abrams, Susan.

Abrams, S. 1995b. "Letter to Professor Thomas S. Kuhn – Dated 20 June 1995," in the TSK Archives. Box 21: Folder 1, Correspondence, Abrams, Susan.

Akers, R. L. 1992. "Linking Sociology and Its Specialties: The Case of Criminology," *Social Forces* 71:1, 1–16.

Almond, G. A. 1966. "Political Theory and Political Science," *American Political Science Review* LX:4, 869–879.

Andersen, H. 2001. *On Kuhn*. Belmont, CA: Wadsworth.

Andersen, Hanne. 2001. "Kuhn, Conant, and Everything – A Full or Fuller Account," *Philosophy of Science* 68:2, 258–262.

Andresen, Jensine. 1999. "Crisis and Kuhn," *Isis* 90 (Supplement), S43–S67.

Arabatzis, Theodore. 2001. "Can a Historian of Science Be a Scientific Realist?," *Philosophy of Science* 68:3 (Supplement: Proceedings of the 2000 Biennial Meeting of the Philosophy of Science Association. Part I: Contributed Papers), S531–S541.

Ayer, A. J. 1959. "Editor's Introduction," in A. J. Ayer (ed.), *Logical Positivism*, pp. 3–28. New York, NY: The Free Press.

Babich, Babette E. 2003. "From Fleck's Denkstil to Kuhn's Paradigms: Conceptual Schemes and Incommensurability," *International Studies in the Philosophy of Science* 17:1, 75–92.

Baltes, P. B., and J. R. Nesselroade. 1984. "Paradigm Lost and Paradigm Regained: Critique of Dannefer's Portrayal of Life-Span Developmental Psychology," *American Sociological Review* 49:6, 841–847.

Barber, B. 1987. "The Emergence and Maturation of the Sociology of Science," *Science & Technology Studies* 5:3/4, 129–133.

Barnes, B. 1974. *Scientific Knowledge and Sociological Theory*. London: Routledge and Kegan Paul.

Barnes, B. 1982. *T. S. Kuhn and Social Science*. New York, NY: Columbia University Press.

Barnes, B., and D. Bloor. 1982. "Relativism, Rationalism and the Sociology of Knowledge," in M. Hollis and S. Lukes (eds.), *Rationality and Relativism*, pp. 21–47. Cambridge, MA: MIT Press.

Barnes, Barry, David Bloor and John Henry. 1996. *Scientific Knowledge: A Sociological Analysis.* Chicago, IL: University of Chicago Press.

Bartlett, P. D. 1983. "James Bryant Conant, 1893–1978: A Biographical Memoir by Paul D. Bartlett," in *National Academy of Sciences, Biographical Memoir*, pp. 89–124. Washington, DC: National Academy of Sciences.

Bensaude-Vincent, Bernadette. 2005. "Chemistry in the French Tradition of Philosophy of Science: Duhem, Meyerson, Metzger and Bachelard," *Studies in History and Philosophy of Science* 36: 627–648.

Biddle, Justin. 2011. "Putting Pragmatism to Work in the Cold War: Science, Technology, and Politics in the Writings of James B. Conant," *Studies in History and Philosophy of Science* 42: 522–561.

Bird, A. 2000. *Thomas Kuhn.* Princeton, NJ: Princeton University Press.

Bird, A. 2015. "Kuhn and the Historiography of Science," in W. J. Devlin and A. Bokulich (eds.), *Kuhn's Structure of Scientific Revolutions – 50 Years On*, pp. 23–38. Dordrecht: Springer.

Birner, J. 2018. "Karl Popper's *The Poverty of Historicism* After 60 Years," *Metascience* 27:2, 183–193.

Blaug, M. 1976. "Kuhn Versus Lakatos, or Paradigms Versus Research Programmes in the History of Economics," in S. J. Latsis (ed.), *Method and Appraisal in Economics*, pp. 149–180. Cambridge: Cambridge University Press.

Bleaney, B. 1982. "John Hasbrouck Van Vleck. 13 March 1899. 27 October 1980," *Biographical Memoirs of Fellows of the Royal Society* 28 (November), 627–665.

Bloor, D. 1975. "Course Bibliography: A Philosophical Approach to Science," *Social Studies of Science* 5: 507–517.

Blumenthal, Geoffrey. 2013. "Kuhn and the Chemical Revolution: A Reassessment," *Foundations of Chemistry* 15: 93–101.

Bohr, Niels. 1948. "On the Notions of Causality and Complementarity," *Dialectica* 2:2–3, 312–319.

Bornmann, L., K. B. Wray and R. Haunschild. 2020. "Citation Concept Analysis (CAA) – A New Form of Citation Analysis Revealing the Usefulness of Concepts for Other Researchers Illustrated by Exemplary Case Studies Including Classic Books by Thomas S. Kuhn and Karl R. Popper," *Scientometrics* 122:2, 1051–1074.

Boyd, Richard N. 1985. "Lex Orandi est Lex Credendi," in Paul M. Churchland and Clifford A. Hooker (eds.), *Images of Science: Essays on Realism and Empiricism, with a Reply from Bas C. van Fraassen*, pp. 3–34. Chicago, IL: University of Chicago Press.

Brasch, Frederick E. 1915, November 26. "The Teaching of the History of Science," *Science* 42:1091, 746–760.

Breiger, R. L. 2005. "White, Harrison," in *Encyclopedia of Social Theory*, pp. 885–886. Thousand Oaks: SAGE Publications, Inc.

Brorson, Stig, and Hanne Andersen. 2001. "Stabilizing and Changing Phenomenal Worlds: Ludwik Fleck and Thomas Kuhn on Scientific Literature," *Journal for General Philosophy of Science* 32:1, 109–129.

Brush, S. G. 2000. "Thomas Kuhn as Historian of Science," *Science & Education* 9, 39–58.

Burian, R. M. 1977. "More than a Marriage of Convenience: On the Inextricability of History and Philosophy of Science," *Philosophy of Science* 44:1, 1–42.

Burian, R. M. 2001. "The Dilemma of Case Studies Resolved: The Virtues of Using Case Studies in the History and Philosophy of Science," *Perspectives on Science* 9:4, 383–404.

Burman, J. T. 2020. "On Kuhn's Case, and Piaget's: A Critical Two-Sited Hauntology (Or, on Impact Without Reference)," *History of the Human Science* 33:3–4, 129–159.

Bush, V. 1946/1975. *Endless Horizons*. New York, NY: Arno Press.

Cammack, P. 1990. "A Critical Assessment of the New Elite Paradigm," *American Sociological Review* 55, 415–420.

Carnap, Rudolf. 1932/1959. "The Elimination of Metaphysics Through Logical Analysis of Language," in A. J. Ayer (ed.), *Logical Positivism*, pp. 60–81. New York, NY: The Free Press.

Carnap, Rudolf. 1956. "The Methodological Status of Theoretical Concepts," in Herbert Feigl and Michael Scriven (eds.), *The Foundations of Science and the Concepts of Psychology and Psychoanalysis*, pp. 38–75. Minneapolis: University of Minnesota Press.

Cavell, S. 2010. *Little Did I Know: Excerpts from Memory*. Stanford, CA: Stanford University Press.

Cedarbaum, Daniel G. 1983. "Paradigms," *Studies in History and Philosophy of Science* 14:3, 173–213.

Chang, Hasok. 2015. "The Chemical Revolution Revisited," *Studies in History and Philosophy of Science* 49, 91–98.

Chang, Hasok, and Nancy Cartwright. 2014. "Measurement," in Martin Curd and Stathis Psillos (eds.), *The Routledge Companion to Philosophy of Science*, 2nd ed., pp. 411–419. London: Routledge.

Coats, A. W. 1969. "Is There a 'Structure of Scientific Revolutions' in Economics?," *Kyklos: International Zeitschrift für Sozialwissenschaften* 22:2, 289–296.

Cohen, I. Bernard. 1985. *Revolution in Science*. Cambridge, MA: The Belknap Press of Harvard University Press.

Cole, J. R., and H. Zuckerman. 1975. "The Emergence of a Scientific Specialty: The Self-Exemplifying Case of the Sociology of Science," in L. A. Coser (ed.), *The Idea of Social Structure: Papers in Honor of Robert K. Merton*, pp. 139–174. New York, NY: Harcourt Brace Jovanovich.

Cole, Stephen. 1992. *Making Science: Between Nature and Society*. Cambridge, MA: Harvard University Press.

Collodel, M. 2016. "Was Feyerabend a Popperian? Methodological Issues in the History of the Philosophy of Science," *Studies in History and Philosophy of Science* 57: 27–56.

Conant, James B. 1946/1947. "Preface," in James B. Conant (ed.) *On Understanding Science: An Historical Approach*. pp. xi–xv. New Haven, CT: Yale University Press.

Conant, James B. 1947. *On Understanding Science: An Historical Approach*. New Haven, CT: Yale University Press.

Conant, James B. 1950/1965. "Foreword," in James B. Conant (ed.), *Robert Boyle's Experiments in Pneumatics: Case 1 in the Harvard Case Histories in Experimental Science*. Cambridge, MA: Harvard University Press.

Conant, James B. 1953. *Modern Science and Modern Man*. Garden City, NY: A Doubleday Anchor Book.

Conant, James B. (ed.) 1955. *The Overthrow of the Phlogiston Theory: The Chemical Revolution of 1775–1789*. Cambridge, MA: Harvard University Press.

Conant, James B. 1957. "Foreword," in T. S. Kuhn (ed.), *The Copernican Revolution: Planetary Astronomy in the Development of Western Thought*, pp. xiii–xviii. Cambridge, MA: Harvard University Press.

Conant, James B. 1961. "Letter from James B. Conant to Kuhn, June 5, 1961," in Thomas S. Kuhn Papers, MC240, Box 25: SSR 1962; Correspondence: Prepublication; Massachusetts Institute of Technology, Institute Archives and Special Collections.

Conant, James B. 1962. "Letter from James B. Conant to Kuhn, December 19, 1962," in Thomas S. Kuhn Papers, MC240, Box 4: Correspondence C-D; Massachusetts Institute of Technology, Institute Archives and Special Collections.

Conant, James B. 1970. *My Several Lives: Memoirs of a Social Inventor*. New York, NY: Harper and Row.

Conant, James and John Haugeland. 2000. "Editors' Introduction," in James Conant and John Haugeland (eds.), Thomas S. Kuhn, *The Road since Structure: Philosophical Essays, 1970–1993, with an Autobiographical Interview*, pp. 1–9. Chicago, IL: University of Chicago Press.

Conant, Jennet. 2017. *Man of the Hour: James B. Conant, Warrior Scientist*. New York, NY: Simon & Schuster.

Creath, R. 2007. "Vienna, the City of Quine's Dreams," in A. Richardson and T. Uebel (eds.), *The Cambridge Companion to Logical Empiricism*, pp. 332–345. Cambridge: Cambridge University Press.

Curd, Martin, J. A. Cover and Christopher Pincock. 2013. *Philosophy of Science: The Central Issues*, 2nd ed. New York, NY: W. W. Norton & Company.

D'Agostino, F. 2005. "Kuhn's Risk-Spreading Argument and the Organization of Scientific Communities," *Episteme* 1:3, 201–209.

Daston, L. 2016. "History of Science without *Structure*," in R. J. Richards and L. Daston (eds.), *Kuhn's Structure of Scientific Revolutions at Fifty: Reflections on a Scientific Classic*, pp. 115–132. Chicago, IL: University of Chicago Press.

Davidson, Donald. 1973–1974. "On the Very Idea of a Conceptual Scheme," *Proceedings and Addresses of the American Philosophical Association* 47: 5–24.

Devlin, William J., and Alisa Bokulich (eds.). 2015. *Kuhn's Structure of Scientific Revolutions – 50 Years On*. Dordrecht: Springer.

Dicken, Paul. 2016. *A Critical Introduction to Scientific Realism*. London: Bloomsbury Academic.

DiSalle, R. 2002. "Reconsidering Kant, Friedman, Logical Positivism, and the Exact Sciences," *Philosophy of Science* 69: 191–211.

Doppelt, Gerald. 1978. "Kuhn's Epistemological Relativism: An Interpretation and Defense," *Inquiry* 21: 33–86.

Doppelt, Gerald. 2013. "Explaining the Success of Science: Kuhn and Scientific Realists," *Topoi* 32: 43–51.

Duhem, Pierre. 1893/1996. "The English School and Physical Theories," in Pierre Duhem, *Essays in the History and Philosophy of Science*, translated and ed. by Roger Ariew and Peter Barker, pp. 50–74. Indianapolis, IN: Hackett Publishing Company.

Duhem, Pierre. 1915/1996. "Some Reflections in German Science," in Pierre Duhem, *Essays in the History and Philosophy of Science*, translated and ed. by Roger Ariew and Peter Barker, pp. 251–276. Indianapolis, IN: Hackett Publishing Company.

Dupree, A. H., and T. S. Kuhn. 1958. "Teaching the History of Science," *Isis* 49:2, 172–173.

Durkheim, É. 1930/1951. *Suicide: A Study in Sociology*, translated by J. A. Spaulding and G. Simpson. New York, NY: The Free Press.

Easlea, B. 1973. "Bibliography: An Introduction to the History and Social Studies of Science: A Seminar Course for First-Year Science Students," *Science Studies* 3: 185–209.

Edge, David. 1967. "Letter to Professor Kuhn – Dated 13 November 1967," in Thomas S. Kuhn Papers, MC240, Box 12: Social Studies of Science, 1970–1976.

Edge, David. 1969. "Letter to Professor Kuhn – Dated 4th December 1969," in Thomas S. Kuhn Papers, MC240, Box 12: Social Studies of Science, 1970–1976.

Eckberg, D. L., and L. Hill, Jr. 1979. "The Paradigm Concept and Sociology: A Critical Review," *American Sociological Review* 44: 925–937.

Edmonds, D. 2020. *The Murder of Professor Schlick: The Rise and Fall of the Vienna Circle*. Princeton, NJ: Princeton University Press.

Edmonds, D., and J. Eidinow. 2002. *Wittgenstein's Poker: The Story of a Ten-Minute Argument Between Two Great Philosophers*. New York, NY: Ecco.

Farley, John, and Gerald L. Geison. 1974. "Science, Politics and Spontaneous Generation in Nineteenth-Century France: The Pasteur–Pouchet Debate," *Bulletin of the History of Medicine* 48: 161–198.

Feyerabend, Paul K. 1958/1981. "An Attempt at a Realistic Interpretation of Experience," in Paul K. Feyerabend, *Realism, Rationalism and Scientific Method: Philosophical Papers, Volume 1*, pp. 17–36. Cambridge: Cambridge University Press.

Feyerabend, Paul K. 1960/1981. "On the Interpretation of Scientific Theories," in Paul K. Feyerabend, *Realism, Rationalism and Scientific Method: Philosophical Papers, Volume 1*, pp. 37–43. Cambridge: Cambridge University Press.

Feyerabend, Paul K. 1964/1981. "Realism and Instrumentalism: Comments on the Logic of Factual Support," in Paul K. Feyerabend, *Realism, Rationalism and*

Scientific Method: Philosophical Papers, Volume 1, pp. 176–202. Cambridge: Cambridge University Press.

Feyerabend, P. 1968/1999. "Letter to Imre Lakatos, Saturday, end of March 1968," in I. Lakatos and P. Feyerabend, *For and Against Method: Including Lakatos's Lectures on Scientific Method and the Lakatos-Feyerabend Correspondence*, edited and with an introduction by M. Motterlini, pp. 141–142. Chicago, IL: University of Chicago Press.

Feyerabend, P. 1969/1999. "Letter to Imre Lakatos, 17 October 1969," in I. Lakatos and P. Feyerabend, *For and Against Method: Including Lakatos's Lectures on Scientific Method and the Lakatos–Feyerabend Correspondence*, edited and with an introduction by M. Motterlini, pp. 180–181. Chicago, IL: University of Chicago Press.

Feyerabend, P. 1970. "Consolations for the Specialist," in I. Lakatos and A. Musgrave (eds.), *Criticism and the Growth of Knowledge: Proceedings of the International Colloquium in the Philosophy of Science, London, 1965, volume 4*, pp. 197–230. Cambridge: Cambridge University Press.

Field, G. L., and J. Higley. 1980. *Elitism*. London: Routledge and Kegan Paul.

Fleck, L. 1935/1979. *Genesis and Development of a Scientific Fact*, edited by T. J. Trenn and R. K. Merton, translated by F. Bradley and T. J. Trenn, with a Foreword by T. S. Kuhn. Chicago, IL: University of Chicago Press.

Forbes, Curtis. 2017. "A Pragmatic Existentialist Approach to the Scientific Realism Debate," *Synthese* 194:9, 3327–3346.

Forrester, John. 2007. "On Kuhn's Case: Psychoanalysis and the Paradigm," *Critical Inquiry* 33: 782–819.

Frank, Philipp. 1949/1961. "Introduction: Historical Background," in Philipp Frank (ed.), *Modern Science and Its Philosophy*, pp. 13–61. New York, NY: Collier Books.

Frank, Philipp. 1952. "Letter to Professor Tom Kuhn – December 2, 1952," in Thomas S. Kuhn Papers, MC240, Box 25: Books: SSR 1962; Correspondence: Pre-publication.

Franklin, Allan. 1986. *The Neglect of Experiment*. Cambridge: Cambridge University Press.

French, R., and M. Gross. 1973. "A Survey of North American Graduate Students in the History of Science 1970–71," *Science Studies* 3: 161–171.

Friedman, M. 2001. *Dynamics of Reason: The 1999 Kant Lectures at Stanford University*. Stanford, CA: Center for the Study of Language and Information Publications.

Friedman, M. 2003. "Kuhn and Logical Empiricism," in T. Nickles (ed.), *Thomas Kuhn*, pp. 19–44. Cambridge: Cambridge University Press.

Fuller, S. 2000. *Thomas Kuhn: A Philosophical History for Our Times*. Chicago, IL: University of Chicago Press.

Fuller, Steve. 2004. *Kuhn vs. Popper: The Struggle for the Soul of Science*. New York, NY: Columbia University Press.

Galison, P. 2016. "Practice All the Way Down," in R. J. Richards and L. Daston (eds.), *Kuhn's Structure of Scientific Revolutions at Fifty: Reflections on a Scientific Classic*, pp. 42–69. Chicago, IL: University of Chicago Press.

Galison, P. 1981. "Kuhn and the Quantum Controversy," *British Journal for the Philosophy of Science* 32:1, 71–85.

Gattei, G. 2008. *Thomas Kuhn's "Linguistic Turn" and the Legacy of Logical Positivism: Incommensurability, Rationality and the Search for Truth*. London: Routledge.

Giere, R. N. 1973. "History and Philosophy of Science: Intimate Relationship or Marriage of Convenience?," *British Journal for the Philosophy of Science* 24:3, 282–297.

Giere, R. 1997. "Kuhn's Legacy for North American Philosophy of Science," *Social Studies of Science*, 27: 496–498.

Giere, Ronald N. 2013. "Kuhn as Perspectival Realist," *Topoi* 32: 53–57.

Gillispie, C. C. 1962. "The Nature of Science: Normal Science Is Succeeded by a Creative Phase of Revolution Out of Which New Concepts Emerge," *Science*, 138 (December 14, 1962), 1251–1253.

Gillispie, C. C. 1993. "Letter to Tom – Dated 31 January 1993," in Thomas S. Kuhn Papers, MC240, Box 21: Folder 35, Correspondence, Gillispie, Charles.

Godfrey-Smith, P. 2003. *Theory and Reality: An Introduction to the Philosophy of Science*. Chicago, IL: University of Chicago Press.

Gordon, Peter E. 2012. "Agonies of the Real: Anti-Realism from Kuhn to Foucault," *Modern Intellectual History* 9:1, 127–147.

Grandy, R. E. 2003. "Kuhn's World Changes," in T. Nickles (ed.), *Thomas Kuhn*, pp. 246–260. Cambridge: Cambridge University Press.

Grene, Marjorie. 1963. "Letter from Marjorie Grene to Kuhn, September 25, 1963," in Thomas S. Kuhn Papers, MC240, Box 4: Folder 9, Correspondence E-G; Massachusetts Institute of Technology, Institute Archives and Special Collections.

Gutting, G. 2003. "Thomas Kuhn and French Philosophy of Science," in T. Nickles (ed.), *Thomas Kuhn*, pp. 45–64. Cambridge: Cambridge University Press.

Hacking, Ian. 1981. "Do We See with a Microscope?," *Pacific Philosophical Quarterly* 62:4, 305–322.

Hacking, Ian. 1983. *Representing and Intervening: Introductory Topics in the Philosophy of Natural Science*. Cambridge: Cambridge University Press.

Hacking, I. 1999. *The Social Construction of What?* Cambridge, MA: Harvard University Press.

Hacking, I. 2012. "Introductory Essay," in T. S. Kuhn, *Structure of Scientific Revolutions*, 4th ed., vii–xxxvii. Chicago, IL: University of Chicago Press.

Hacking, I. 2016. "Paradigms," in Robert J. Richards and Lorraine Daston (eds.), *Kuhn's Structure of Scientific Revolutions at Fifty: Reflections on a Science Classic*, 96–112. Chicago, IL: University of Chicago Press.

Hacohen, Malachi H. 2002. *Karl Popper – The Formative Years, 1902–1945: Politics and Philosophy in Interwar Vienna*. Cambridge: Cambridge University Press.

Hall, Richard J. 1970. "Kuhn and the Copernican Revolution," *British Journal for the Philosophy of Science* 21:2, 196–197.

Hallberg, M. 2017. "Revolutions and Reconstructions in the Philosophy of Science: Mary Hesse (1924–2016)," *Journal for General Philosophy of Science* 48: 161–171.

Hamlin, Christopher. 2016. "The Pedagogical Roots of the History of Science: Revisiting the Vision of James Bryant Conant." *Isis* 107:2, 282–308.

Hayek, F. 1964. "The Theory of Complex Phenomena," in M. A. Bunge (ed.), *The Critical Approach to Science and Philosophy*, pp. 332–349. Glencoe, NY: Free Press.

Heilbron, J. L. 1968. "Quantum Historiography and the Archive for History of Quantum Physics," *History of Science* 7: 90–111.

Heilbron, J. L. 1998. "Thomas Samuel Kuhn, 18 July 1922 – 17 June 1996." *Isis* 89:3, 505–515.

Herring, Pendleton. 1959. "Letter to Professor Thomas Kuhn – Dated December 7, 1959," in Thomas S. Kuhn Papers, MC240, Box 23: Lecture/Meetings: Berkeley Social Science Colloquium, 1956.

Hershberg, James G. 1993. *James B. Conant: Harvard to Hiroshima and the Making of the Nuclear Age*. New York, NY: Alfred A. Knopf.

Herschbach, D. R. 2014. "Theodore William Richards: Apostle of Atomic Weights and Nobel Prize Winner in 1914," *Angewandte Chemie* 53: 13,982–13,987.

Hesse, Mary 1963. "Review of *The Structure of Scientific Revolutions* by Thomas S. Kuhn," *Isis* 54:2, 286–287.

Hesse, Mary. 1976. "Truth and the Growth of Scientific Knowledge," *PSA: Proceedings of the Biennial Meeting of the Philosophy of Science Association*, pp. 261–280.

Hesse, Mary. 1980. *Revolutions and Reconstructions in the Philosophy of Science*. Brighton: The Harvester Press.

Hesse, M. 1982a. "Comments on the Papers of David Bloor and Steven Lukes," *Studies in History and Philosophy of Science* 4: 325–331.

Hesse, M. 1982b. "Comment on Kuhn's 'Commensurability, Comparability, and Communicability,'" *PSA: Proceedings of the Biennial Meeting of the Philosophy of Science Association*, 1982, *Vol. Two: Symposia and Invited Papers (1982)*, pp. 704–711.

Hesse, Mary. 2002. Hesse, Mary (1924–2016), Text of speech made on 21/6/02. Held at Cambridge, Whipple Library. Hesse Papers, MH2.22.

Howard, D. 2004. "Who Invented the 'Copenhagen Interpretation'? A Study in Mythology," *Philosophy of Science* 71:5 (Proceedings of the 2002 Biennial Meeting of the PSA, Part II: Symposia Papers), 669–682.

Hoyningen-Huene, P. 1989/1993. *Reconstructing Scientific Revolutions: Thomas S. Kuhn's Philosophy of Science*, translated by A. T. Levine, with a foreword by T. S. Kuhn. Chicago, IL: University of Chicago Press.

Hoyningen-Huene, P. 1992. "The Interrelations between the Philosophy, History and Sociology of Science in Thomas Kuhn's Theory of Scientific Development," *British Journal for the Philosophy of Science* 43, 487–501.

Hoyningen-Huene, Paul. 2005. "Three Biographies: Kuhn, Feyerabend, and Incommensurability," in R. A Harris (ed.), *Rhetoric and Incommensurability*, pp. 150–175. West Lafayette, IN: Parlor Press.

Hoyningen-Huene, Paul. 2006. "More Letters from Paul Feyerabend to Thomas S. Kuhn on *Proto-Structure*," *Studies in History and Philosophy of Science* 37, 610–632.

Hoyningen-Huene, Paul. 2008. "Thomas Kuhn and the Chemical Revolution," *Foundations of Chemistry* 10: 101–105.

Hoyningen-Huene, P. 2015. "Kuhn's Development Before and After *Structure*," in W. J. Devlin and A. Bokulich (eds.), *Kuhn's Structure of Scientific Revolutions – 50 Years On*. Boston Studies in the Philosophy and History of Science, Vol. 311, pp. 185–195. Dordrecht: Springer.

Hoyningen-Huene, P., and H. Sankey (eds.). 2001. *Incommensurability and Related Matters*, Vol. 216, Boston Studies in the Philosophy and History of Science. Dordrecht: Springer Netherlands.

Hufbauer, Karl. 2012. "From Student of Physics to Historian of Science: T. S. Kuhn's Education and Early Career, 1940–1958," *Physics in Perspective*, 14: 421–470.

Irzik, G. 2012. "Kuhn and Logical Positivism: Gaps, Silences, and Tactics of SSR," in V. Kindi and T. Arabatzis (eds.), *Kuhn's The Structure of Scientific Revolutions Revisited*, pp. 15–40. London: Routledge.

Isaac, Joel. 2012. *Working Knowledge: Making the Human Sciences from Parsons to Kuhn*. Cambridge, MA: Harvard University Press.

Jacobs, Struan. 2007. "Michael Polanyi and Thomas Kuhn: Priority and Credit," *Tradition and Discovery*, 33:2, 25–36.

Jacobs, Struan. 2010. "J. B. Conant's Other Assistant: Science as Depicted by Leonard K. Nash, Including Reference to Thomas Kuhn," *Perspectives on Science* 18:3, 328–351.

James, William. 1907/1949. *Pragmatism: A New Name for Some Old Ways of Thinking, Together with Four Related Essays Selected from The Meaning of Truth*. New York, NY: Longmans Green and Co.

Joergensen, Joergen. 1970. "The Development of Logical Empiricism," in Otto Neurath, Rudolf Carnap and Charles Morris (eds.), *Foundations of the Unity of Science: Toward an International Encyclopedia of Unified Science, Volume 2*, pp. 845–946. Chicago, IL: University of Chicago Press.

Kahneman, D. 2003. "Maps of Bounded Rationality: Psychology for Behavioral Economics," *American Economic Review* 93:5, 1449–1475.

Kaiser, D. 2016. "Thomas Kuhn and the Psychology of Scientific Revolutions," in R. J. Richards and L. Daston (eds.), *Kuhn's Structure of Scientific Revolutions at Fifty: Reflections on a Scientific Classic*, pp. 71–95. Chicago, IL: University of Chicago Press.

Kaiserlian, P. 1986. "Letter to Professor Thomas Kuhn – June 12, 1986," in Thomas S. Kuhn Papers, MC240, Box 25: Books: SSR 1962; Correspondence: Pre-publication.

Kevles, D. J. 1977. "The National Science Foundation and the Debate over Postwar Research Policy, 1942–1945: A Political Reinterpretation of *Science – The Endless Frontier*," *Isis* 68:1, 4–26.

Kindi, Vasso. 2017. "Wittgenstein and Philosophy of Science," in Hans-Johann Glock and John Hyman (eds.), *A Companion to Wittgenstein*, pp. 587–602. Oxford: Wiley-Blackwell.

Kindi, Vasso, and Theodore Arabatzis (eds.). 2012. *Kuhn's The Structure of Scientific Revolutions Revisited*. London: Routledge.

Klein, M. J. 1979. "Paradigm Lost? A Review Symposium," *Isis* 70:253, 430–434.

Kosiński, J. 1970. *Being There*. London: The Bodley Head.

Kourany, Janet A. 1998. *Scientific Knowledge: Basic Issues in the Philosophy of Science*, 2nd ed. Belmont, CA: Wadsworth Publishing Company.

Kuhn, Thomas S. 1949. "Notebook" in Thomas S. Kuhn Papers, MC240, Box 1: Folder 7, Notes and Ideas, 1949; Massachusetts Institute of Technology, Institute Archives and Special Collections.

Kuhn, Thomas S. 1951a. "Newton's '31st Query' and the Degradation of Gold." *Isis*, 42:4, 296–298.

Kuhn, T. S. 1951b. "The Quest for Physical Theory," in Thomas S. Kuhn Papers, MC240, Box 3, Lectures, Lowell Institute, 1950–51.

Kuhn, Thomas S. 1952. "Robert Boyle and Structural Chemistry in the Seventeenth Century." *Isis* 43:1, 12–36.

Kuhn, Thomas S. 1953. "Letter to Charles Morris – Dated July 31, 1953," in Thomas S. Kuhn Papers, MC240, Box 25: Books: The Structure of Scientific Revolutions (1962) – Correspondence, pre-publication, 1952–1962.

Kuhn, Thomas. S. 1956/1957. "Preface," in Thomas S. Kuhn, *The Copernican Revolution: Planetary Astronomy in the Development of Western Thought*, pp. vii–x. Cambridge, MA: Harvard University Press.

Kuhn, Thomas. S. 1957. *The Copernican Revolution: Planetary Astronomy in the Development of Western Thought*. Cambridge, MA: Harvard University Press.

Kuhn, Thomas S. 1958. "Letter to Professor R. K. Merton – Dated September 16, 1958," in Thomas S. Kuhn Papers, MC240, Box 22: Correspondence: Merton, Robert K.

Kuhn, T. S. 1959a. "Letter to Ralph W. Tyler, Director – Dated July 31, 1959," in Thomas S. Kuhn Papers, MC240, Box 14: Folder 15, Center for Advanced Study, 1959.

Kuhn, T. S. 1959b. "Letter to Dr. Pendleton Herring – Dated 21 December 1959," in Thomas S. Kuhn Papers, MC240, Box 14: Folder 15, Center for Advanced Study, 1959.

Kuhn, Thomas S. 1959c. "Letter to Professor Robert K. Merton – Dated 7 December 1959," in Thomas S. Kuhn Papers, MC240, Box 23: Lecture/Meetings: Berkeley Social Science Colloquium, 1956.

Kuhn, T. S. 1959/1977. "Energy Conservation as an Example of Simultaneous Discovery," in *Essential Tension: Selected Studies in Scientific Tradition and Change*, pp. 66–104. Chicago, IL: University of Chicago Press.

Kuhn, T. S. 1961a. "Letter to Mr. Carroll G. Bowen – Dated 18 June 1961," in Thomas S. Kuhn Papers, MC240, Box 25: Books: SSR 1962; Correspondence: Pre-publication.

Kuhn, T. S. 1961b. "Letter to Dr. James B. Conant – Dated 5 August 1961," in Thomas S. Kuhn Papers, MC240, Box 25: Books: SSR 1962; Correspondence: Pre-publication.

Kuhn, Thomas S. 1961/1977. "The Function of Measurement in Modern Physical Science," in Thomas Kuhn, *Essential Tension: Selected Studies in Scientific Tradition and Change*, pp. 178–224. Chicago, IL: University of Chicago Press.

Kuhn, T. S. 1962/2012. *The Structure of Scientific Revolutions*, 4th ed., with an introductory essay by Ian Hacking. Chicago, IL: University of Chicago Press.

Kuhn, Thomas S. 1962. "Historical Structure of Scientific Discovery," *Science* 136:3518 (June 1, 1962), pp. 760–764.

Kuhn, T. S. 1963. "Letter to George Stigler – Dated 24 October 1963," in Thomas S. Kuhn Papers, MC240, Box 4: Folder 15, Structure of Scientific Revolutions, Correspondence S. Institute Archives and Special Collections, MIT Libraries, Cambridge, Massachusetts.

Kuhn, T. S. 1966a. "Letter to Miss Margaret Masterman – Dated 1 June 1966," in Thomas S. Kuhn Papers, MC240, Box 11: Correspondence: Masterman, Margaret.

Kuhn, T, S. 1966b. "Lecture Notes: Phil – Hum 537, Meeting #1 – 2/11/66," in Thomas S. Kuhn Papers, MC240, Box 2: Folder 10, Development of Scientific Knowledge 537: 1966.

Kuhn, T. S. 1968/1977. "The History of Science," in T. S. Kuhn, *Essential Tension: Selected Studies in Scientific Tradition and Change*, pp. 105–126. Chicago, IL: University of Chicago Press.

Kuhn, T. S. 1969/1977. "Comments on the Relations of Science and Art," in T. S. Kuhn, *The Essential Tension: Selected Studies in Scientific Tradition and Change*, pp. 340–351. Chicago, IL: University of Chicago Press.

Kuhn, Thomas. S. 1969/2012. "Postscript – 1969," in *Structure of Scientific Revolutions*, 4th ed., with an introductory essay by Ian Hacking, pp. 173–208. Chicago, IL: University of Chicago Press.

Kuhn, T. S. 1970/1977. "Logic of Discovery or Psychology of Research?," in T. S. Kuhn, *The Essential Tension: Selected Studies in Scientific Tradition and Change*, pp. 266–292. Chicago: University of Chicago Press.

Kuhn, T. S. 1970/2000. "Reflections on My Critics," in T. S. Kuhn, *The Road since Structure: Philosophical Essays, 1970–1993, with an Autobiographical Interview*, edited by J. Conant and J. Haugeland, pp. 123–175. Chicago, IL: University of Chicago Press.

Kuhn, T. S. 1971/1977. "The Relations between History and the History of Science," in T. S. Kuhn, *The Essential Tension: Selected Studies in Scientific Tradition and Change*, pp. 127–161. Chicago, IL: University of Chicago Press.

Kuhn, T. S. 1972. "Letter to Professor John L. Heilbron – 5 September 1972," in Thomas S. Kuhn Papers, MC240, Box 21: Folder 40, Correspondence: Heilbron, John.

Kuhn, T. S. 1973. "Letter to Professor Kandall – 20 June 1973," in Thomas S. Kuhn Papers, MC240, Box 4: Folder 11, Structure of Scientific Revolutions, Correspondence K–L.

Kuhn, Thomas S. 1973/1977. "Objectivity, Value Judgment, and Theory Choice," in T. S. Kuhn, *The Essential Tension: Selected Studies in Scientific Tradition and Change*, pp. 320–339. Chicago, IL: University of Chicago Press.

Kuhn, T. S. 1974/1977. "Second Thoughts on Paradigms," in T. S. Kuhn, *The Essential Tension: Selected Studies in Scientific Tradition and Change*, pp. 293–319. Chicago, IL: University of Chicago Press.

Kuhn, Thomas S. 1976. "Letter to David Edge – Dated 21 July 1976," in Thomas S. Kuhn Papers, MC 240, Box 12: Social Studies of Science, 1970–1976.

Kuhn, T. S. 1976/1977a. "The Relations Between the History and the Philosophy of Science," in T. S. Kuhn, *The Essential Tension: Selected Studies in Scientific Tradition and Change*, pp. 3–20. Chicago, IL: University of Chicago Press.

Kuhn, T. S. 1976/1977b. "Mathematical versus Experimental Traditions in the Development of Physical Science," in T. S. Kuhn, *The Essential Tension: Selected Studies in Scientific Tradition and Change*, pp. 31–65. Chicago, IL: University of Chicago Press.

Kuhn, T. S. 1977. "Preface," in T. S. Kuhn, *The Essential Tension: Selected Studies in Scientific Tradition and Change*, pp. ix–xxiii. Chicago, IL: University of Chicago Press.

Kuhn, Thomas. S. 1978/1987. *Black-Body Theory and the Quantum Discontinuity, 1894–1912*. Chicago, IL: University of Chicago Press.

Kuhn, T. S. 1979. "Foreword," in L. Fleck, *Genesis and Development of a Scientific Fact*, edited by T. J. Trenn and R. K. Merton, translated by F. Bradley and T. J. Trenn, pp. vii–xi. Chicago, IL: University of Chicago Press.

Kuhn, Thomas S. 1979/2000. "Metaphor in Science," in Thomas S. Kuhn, *The Road since Structure: Philosophical Essays, 1970–1993, with an Autobiographical Interview*, edited by James Conant and John Haugeland, pp. 196–207. Chicago, IL: University of Chicago Press.

Kuhn, Thomas S. 1983a. "Reflections on Receiving the John Desmond Bernal Award," *4S Review* 1:4, 26–30.

Kuhn, T. S. 1983b. "Letter to Professor Ernan McMullin – Dated March 29, 1983," in Thomas S. Kuhn Papers, MC240, Box 22: Correspondence: McMullin, Ernan.

Kuhn, Thomas S. 1984a. "Professionalization Recollected in Tranquility," *Isis* 75 (1): 29–32.

Kuhn, Thomas S. 1984b. "Scientific Development and Lexical Change: The Thalheimer Lectures," in Thomas S. Kuhn Papers, MC240, Box 23: Lectures/ Meetings: Thalheimer Lectures, Scientific Development and Lexical Change, [2 folders] 1984.

Kuhn, T. S. 1986. "Letter to Kenneth Grau – Dated 24 June 1986," in Thomas S. Kuhn Papers, MC240, Box 20: Folder 16, Advice, 1986–1987.

Kuhn, T. S. 1987/2000. "What Are Scientific Revolutions?," in T. S. Kuhn, *The Road since Structure: Philosophical Essays, 1970–1993, with an Autobiographical*

Interview, edited by J. Conant and J. Haugeland, pp. 13–32. Chicago, IL: University of Chicago Press.

Kuhn, T. S. 1989. "Proposal to the National Science Foundation: 1 August 1989," in Thomas S. Kuhn Papers, MC240, Box 20: Folder 12, NSF Research Reports (1 of 2 files).

Kuhn, Thomas S. 1989/2000. "Possible Worlds in the History of Science," in Thomas S. Kuhn, *The Road since Structure: Philosophical Essays, 1970–1993, with an Autobiographical Interview*, edited by James Conant and John Haugeland, pp. 58–89. Chicago, IL: University of Chicago Press.

Kuhn, T. S. 1990/2016. "The Nature of Scientific Knowledge: An Interview with Thomas S. Kuhn," in A. Blum, K. Gavroglu, C. Joas and J. Renn (eds.), *Shifting Paradigms: Thomas S. Kuhn and the History of Science*, pp. 17–30. Edition Open Access, Max Planck Institute for the History of Science. Berlin: Neopubli GmbH.

Kuhn, T. S. 1991. "Letter to Jigang Wei – Dated 9 September 1991," in Thomas S. Kuhn Papers, MC240, Box 20: Folder 40, Correspondence, W.

Kuhn, Thomas S. 1991/2000a. "The Road since *Structure*," in Thomas S. Kuhn, *The Road since Structure: Philosophical Essays, 1970–1993, with an Autobiographical Interview*, edited by James Conant and John Haugeland, pp. 90–104. Chicago, IL: University of Chicago Press.

Kuhn, Thomas S. 1991/2000b. "The Natural and the Human Sciences," in Thomas S. Kuhn, *The Road since Structure: Philosophical Essays, 1970–1993, with an Autobiographical Interview*, edited by James Conant and John Haugeland, pp. 216–223. Chicago, IL: University of Chicago Press.

Kuhn, T. S. 1992. "Letter to Mr. Frederick L. Whitaker – Dated 15 January 1992," in Thomas S. Kuhn Papers, MC240, Box 20: Folder 40, Correspondence, W.

Kuhn, T. S. 1992/2000. "The Trouble with the Historical Philosophy of Science," in T. S. Kuhn's *The Road since Structure: Philosophical Essays, 1970–1993, with an Autobiographical Interview*, edited by J. Conant and J. Haugeland, pp. 105–120. Chicago, IL: University of Chicago Press.

Kuhn, T. S. 1993a. "Letter to Professor J. van Brakel – Dated 30 May 1993," in Thomas S. Kuhn Papers, MC240, Box 20: Folder 39, Correspondence, V.

Kuhn, T. S. 1993b. "Letter to Professor C.C. Gillispie – Dated 13 February 1993," in Thomas S. Kuhn Papers, MC240, Box 21: Folder 35, Correspondence: Gillispie, Charles.

Kuhn, Thomas S. 1997/2000. "A Discussion with Thomas S. Kuhn," in *The Road since Structure: Philosophical Essays. 1970–1993, with an Autobiographical Interview*, edited by James Conant and John Haugeland, pp. 255–323. Chicago, IL: University of Chicago Press.

Kuhn, T. S., J. L. Heilbron, P. Forman and L. Allen. 1967. *Sources for History of Quantum Physics*. American Philosophical Society, Memoir 68. https://amphilsoc.org/guides/ahqp/index.htm (accessed July 29, 2020).

Kusch, Martin. 2015. "Scientific Pluralism and the Chemical Revolution," *Studies in History and Philosophy of Science*, 49: 69–79.

Ladyman, J. 1998. "What Is Structural Realism?," *Studies in History and Philosophy of Science* 29:3, 409–424.

Lakatos, Imre. 1970/1972. "Falsification and the Methodology of Scientific Research Programmes," in Imre Lakatos and Alan Musgrave (eds.), *Criticism and the Growth of Knowledge: Proceedings of the International Colloquium in the Philosophy of Science, London, 1965, Volume 4*, pp. 91–196. Cambridge: Cambridge University Press.

Latour, Bruno. 1987. *Science in Action: How to Follow Scientists and Engineers Through Society*. Cambridge, MA: Harvard University Press.

Latour, Bruno, and Steve Woolgar. 1979/1986. *Laboratory Life: The Construction of Scientific Facts*. Princeton, NJ: Princeton University Press.

Laudan, Larry. 1977. *Progress and Its Problems: Toward a Theory of Scientific Growth*. Berkeley and Los Angeles, CA: University of California Press.

Laudan, Larry. 1981. "Confutation of Convergent Realism," *Philosophy of Science* 48, 19–49.

Laudan, L. 1984. *Science and Values: The Aims of Science and their Role in Scientific Debate*. Berkeley and Los Angeles, CA: University of California Press.

Leijonhufvud, A. 1976. "Schools, 'Revolutions,' and Research Programmes in Economic Theories," in S. J. Latsis (ed.), *Method and Appraisal in Economics*, pp. 65–108. Cambridge: Cambridge University Press.

Leontief, W. 1964. "Letter to Thomas Kuhn," in Thomas S. Kuhn Papers, MC240, Box 4: Folder 11, Structure of Scientific Revolutions, Correspondence K-L. Institute Archives and Special Collections, MIT Libraries, Cambridge, Massachusetts.

Little, D. 1991. *Varieties of Social Explanation: An Introduction to the Philosophy of Social Science*. Boulder, CO: Westview Press, Inc.

Longino, H. E. 1990. *Science as Social Knowledge: Values and Objectivity in Scientific Inquiry*. Princeton, NJ: Princeton University Press.

Mach, Ernst. 1893/1960. *The Science of Mechanics: A Critical and Historical Account of Its Development*, translated by T. J. McCormack, 6th ed. with revisions through the ninth German edition. La Salle, IL: The Open Court Publishing Company.

Marcionis, J. J. 1997. *Sociology*, 6th ed. Upper Saddle River, NJ: Prentice Hall.

Marcum, James A. 2015. *Thomas Kuhn's Revolutions: A Historical and an Evolutionary Philosophy of Science?* London: Bloomsbury.

Martin, P. S. 1971. "The Revolution in Archaeology," *American Antiquity* 36:1, 1–8.

Massimi, Michela. 2015. "Walking the Line: Kuhn Between Realism and Relativism," in William J. Devlin and Alisa Bokulich (eds.), *Kuhn's Structure of Scientific Revolutions – 50 Years On*. Boston Studies in the Philosophy and History of Science 311, pp. 135–152. Dordrecht: Springer International Publishing.

Masterman, M. 1966. "Letter to Professor T. S. Kuhn – 8 June 1966," in Thomas S. Kuhn Papers, MC240, Box 11: Correspondence: Masterman, Margaret.

Masterman, M. 1970/1972. "The Nature of a Paradigm," in I. Lakatos and A. Musgrave (eds.), *Criticism and the Growth of Knowledge: Proceedings of the International Colloquium in the Philosophy of Science, London, 1965*, Vol. IV, reprinted with corrections, pp. 59–89. Cambridge: Cambridge University Press.

Matthews, M. R. 2003. "Thomas Kuhn's Impact on Science Education: What Lessons Can Be Learned?," *Science Education* 88:1, 90–118.

Mauskopf, Seymour H. 2012. "Thomas S. Kuhn and the Chemical Revolution," *Historical Studies in the Natural Sciences*, 42:5, 551–556.

Maxwell, Grover. 1962. "The Ontological Status of Theoretical Entities," in Herbert Feigl and Grover Maxwell (eds.), *Scientific Explanation, Space and Time*, pp. 3–27. Minneapolis: University of Minnesota Press.

Mayr, E. 1994. "The Advance of Science and Scientific Revolutions," *Journal of the History of the Behavioral Sciences* 30, 328–334.

McEvoy, John G. 2010/2016. *The Historiography of the Chemical Revolution*. Routledge: London.

McMichael, Alan. 1985. "Van Fraassen's Instrumentalism," *British Journal for the Philosophy of Science* 36, 257–272.

McMullin E. 1976. "History and Philosophy of Science: A Marriage of Convenience?," in R. S. Cohen, C. A. Hooker, A. C. Michalos and J. W. Van Evra (eds.), *PSA 1974. Boston Studies in the Philosophy of Science*, Vol. 32, pp. 585–681. Dordrecht: Springer.

McMullin, Ernan. 1982. "Values in Science," *PSA: Proceedings of the Biennial Meeting of the Philosophy of Science Association*, 1982, Vol. 1982, Volume Two: Symposia and Invited Papers (1982), 3–28.

McMullin, Ernan. 1992. "Rationality and Paradigm Change in Science," in Paul Horwich (ed.), *World Changes: Thomas Kuhn and the Nature of Science*, pp. 55–78. Cambridge, MA: The MIT Press.

Melogno, P., and A. Courtoisie. 2019. "Stepping into the 60s: Tomas [sic] Kuhn's Intellectual Turn towards the Philosophy of Science," *Daimon: Revista Internacional de Filosofía* 76: 23–33.

Meltzer, D. J. 1979. "Paradigms and the Nature of Change in American Archaeology," *American Antiquity* 44:4, 644–657.

Merton, R. K. 1949/1996. "Manifest and Latent Functions," in R. K. Merton, *On Social Structure and Science*, edited by P. Sztompka, pp. 87–95. Chicago, IL: University of Chicago Press.

Merton, R. K. 1957/1973. "Priorities in Scientific Discovery," in R. K. Merton, *The Sociology of Science: Theoretical and Empirical Investigations*, edited by N. W. Storer, pp. 286–324. Chicago, IL: University of Chicago Press.

Merton, Robert K. 1959. "Letter to Tom – Dated 25 May 1959," in Thomas S. Kuhn Papers, MC240, Box 23: Lecture/Meetings: Berkeley Social Science Colloquium, 1956.

Merton, R. K. 1961/1973. "Singletons and Multiples in Science," in R. K. Merton, *The Sociology of Science: Theoretical and Empirical Investigations*, edited by N. W. Storer, pp. 343–370. Chicago, IL: University of Chicago Press.

Merton, R. K. 1963/1973. "Multiple Discoveries as a Strategic Research Site," in R. K. Merton, *The Sociology of Science: Theoretical and Empirical Investigations*, edited by N. W. Storer, pp. 371–382. Chicago, IL: University of Chicago Press.

Merton, R. K. 1968/1973. "The Matthew Effect in Science," in R. K. Merton, *The Sociology of Science: Theoretical and Empirical Investigations*, edited by N. W. Storer, pp. 439–459. Chicago, IL: University of Chicago Press.

Merton, R. K. 1975. "Letter to Professor Thomas S. Kuhn – Dated 16 April 1975," in Thomas S. Kuhn Papers, MC240, Box 22: Correspondence: Merton, Robert K.

Merton, R. K. 1976. "Letter to Professor Thomas S. Kuhn – Dated 18 February 1976," in Thomas S. Kuhn Papers, MC240, Box 22: Correspondence: Merton, Robert K.

Merton, R. K. 1977. *The Sociology of Science: An Episodic Memoir*. Carbondale and Edwardsville, IL: Southern Illinois University Press.

Merton, R. K. 1988. "The Matthew Effect in Science, II: Cumulative Advantage and the Symbolism of Intellectual Property," *Isis* 79:4, 606–623.

Merton, R. K. and E. Barber. 2004. *The Travels and Adventures of Serendipity*. Princeton, NJ: Princeton University Press.

Mirowski, Philip. 2005. "Hoedown at the OK Corral: More Reflections on the 'Social' in Current Philosophy of Science," *Studies in History and Philosophy of Science* 36, 790–800.

Mizrahi, M. 2020. "The Case Study Method in Philosophy of Science: An Empirical Study," *Perspectives on Science* 28:1, 63–88.

Mladenović, B. 2017. *Kuhn's Legacy: Epistemology, Metaphilosophy, Pragmatism*. New York, NY: Columbia University.

Moleski, Martin X. 2007. "Polanyi vs. Kuhn: Worldviews Apart," *Tradition and Discovery* 33:2, 8–24.

Mößner, Nicola. 2011. "Thought Styles and Paradigms – A Comparative Study of Ludwik Fleck and Thomas S. Kuhn," *Studies in History of Philosophy of Science*, 42:2, 362–371.

Mullins, Phil. 2002. "On Persons and Knowledge: Marjorie Grene and Michael Polanyi," in R. E. Auxier and L. E. Hahn (eds.), *The Philosophy of Marjorie Grene*, pp. 31–60. Chicago, IL: Open Court.

Mullins, Phil. 2009/2010. "In Memoriam: Marjorie Grene," *Tradition & Discovery: The Polanyi Society Periodical* 36:1, 55–69.

Nagel, E. 1936a. "Impressions and Appraisals of Analytic Philosophy in Europe, I," *Journal of Philosophy* 33:1, 5–24.

Nagel, E. 1936b. "Impressions and Appraisals of Analytic Philosophy in Europe, II," *Journal of Philosophy* 33:2, 29–53.

Nagel, E. 1939. *Principles of the Theory of Probability*, 1:6, of International Encyclopedia of Unified Science. Chicago, IL: University of Chicago Press.

Nagel, Ernest. 1961. *The Structure of Science: Problems in the Logic of Scientific Explanation*. London: Routledge & Kegan Paul.

Nash, Leonard K. 1950/1957. "The Atomic-Molecular Theory," in James B. Conant and Leonard K. Nash (ed.), *Harvard Case Histories in Experimental Science*, Vol. I, pp. 215–321. Cambridge, MA: Harvard University Press.

Nash, Leonard K. 1957. "Letter to Thomas Kuhn – Dated 10 April 1957," in Thomas S. Kuhn Papers, MC240, Box 11: Folder 48, Correspondence with Individuals, Nash, Leonard, 1957–63.

Nash, Leonard K. 1963. "Letter to Professor Thomas S. Kuhn – Dated May 24, 1963," in Thomas S. Kuhn Papers, MC240, Box 11: Folder 48, Correspondence with Individuals, Nash, Leonard, 1957–63.

Neubarth, N. L., Alan J. Emanuel, Yin Liu et al. 2020. "Meissner Corpuscles and their Spatial Intermingled Afferents Underlie Gentle Touch Perception," *Science* 368 (June 19, 2020).

Neuber, Matthias. 2011. "Feigl's 'Scientific Realism,'" *Philosophy of Science* 78:1, 165–183.

Nickles, T. 2003. "Normal Science: From Logic to Case-Based and Model-Based Reasoning," in T. Nickles (ed.), *Thomas Kuhn*, pp. 142–177. Cambridge: Cambridge University Press.

Nobel. 1977. "Philip W. Anderson: Biographical," The Nobel Prize. www .nobelprize.org/prizes/physics/1977/anderson/biographical/ (accessed April 1, 2020).

Nola, R. 2000. "Saving Kuhn from the Sociologists of Science," *Science & Education*, 9, 77–90.

NSF. 1989. "History and Philosophy of Science & Technology – Panel Summary," in the TSK Archives. Box 20: Folder 13, NSF (2 of 2).

Nye, M. J. 2011. *Michael Polanyi and His Generation: Origins of the Social Construction of Science.* Chicago, IL: University of Chicago Press.

Nye, M. J. 2019. "Shifting Trends in Modern Physics, Nobel Recognition, and the Histories that We Write," *Physics in Perspective* 21, 3–22.

Oberheim, Eric. 2005. "On the Historical Origins of the Contemporary Notion of Incommensurability: Paul Feyerabend's Assault on Epistemic Conservativism." *Studies in History and Philosophy of Science*, 36:2, 363–390.

Oberheim, E. and P. Hoyningen-Huene. 2018. "The Incommensurability of Scientific Theories," in Edward N. Zalta (ed.), *The Stanford Encyclopedia of Philosophy* (Fall 2018 edition), https://plato.stanford.edu/archives/fall2018/entries/incommensurability/ (accessed February 11, 2021).

Oppenheim, Paul and Hilary Putnam. 1958. "Unity of Science as a Working Hypothesis," in Herbert Feigl, Michael Scriven and Grover Maxwell (eds.), *Concepts, Theories, and the Mind-Body Problem*, pp. 3–36. Minneapolis: University of Minnesota Press.

Papineau, D. 1974. "Review of the Structure of Scientific Inference by Mary Hesse," *Cambridge Review*, May 1974, 167–168.

Patton, L. 2018. "Kuhn, Pedagogy, and Practice: A Local Reading of *Structure*," in M. Mizrahi (ed.), *The Kuhnian Image of Science: Time for a Decisive Transformation?*, pp. 113–130. London: Rowman & Littlefield International Ltd.

Pinch, T. J. 1979. "Paradigm Lost? A Review Symposium," *Isis* 70:253, 437–440.

Pinto de Oliveira, J. C. 2017. "Thomas Kuhn, the Image of Science and the Image of Art: The First Manuscript of *Structure*," *Perspectives on Science* 25:6, 746–765.

Pitt, J. C. 2001. "The Dilemma of Case Studies: Toward a Heraclitian Philosophy of Science," *Perspectives on Science* 9:4, 373–382.

Planck, Max. 1909/1992. "The Unity in the Physical World Picture," in J. Blackmore (ed.), *Ernst Mach – A Deeper Look: Documents and New Perspectives*, pp. 141–146. Dordrecht: Kluwer Academic Publishers.

Poincaré, H. 1903/2001. "Science and Hypotheses," in S.J. Gould (ed.), *The Value of Science: Essential Writings of Henri Poincaré*. New York, NY: The Modern Library.

Polanyi, Michael. 1962. *Personal Knowledge: Towards a Post-Critical Philosophy*. Chicago, IL: University of Chicago Press.

Politi, V. 2018. "Scientific Revolutions, Specialization and the Discovery of the Structure of DNA: Toward a New Picture of the Development of the Sciences," *Synthese* 195: 2267–2293.

Polsby, N. W. 1998. "Social Science and Scientific Change: A Note on T. S. Kuhn's Contribution," *Annual Review of Political Science* 1, 199–210.

Popper, Karl R. 1935/1992. *The Logic of Scientific Discovery*. London: Routledge.

Popper, K. 1944. "The Poverty of Historicism, I," *Economica*, 11:42 (May 1944), 86–103.

Popper, Karl R. 1956/1963. "Three Views Concerning Human Knowledge," in Karl R. Popper, *Conjectures and Refutations: The Growth of Scientific Knowledge*, pp. 130–160. London: Routledge.

Popper, K. 1957/1991. *The Poverty of Historicism*. London: Routledge.

Popper, Karl R. 1963. "Science: Conjectures and Refutations," in Karl R. Popper, *Conjectures and Refutations: The Growth of Scientific Knowledge*, pp. 43–86. London: Routledge.

Popper, K. R. 1970/1972. "Normal Science and Its Dangers," in I. Lakatos and A. Musgrave (eds.), *Criticism and the Growth of Knowledge: Proceedings of the International Colloquium in the Philosophy of Science, London, 1965*, Vol. IV, reprinted with corrections, pp. 51–58. Cambridge: Cambridge University Press.

Popper, Karl R. 1974. "Replies to My Critics," in Paul A. Schilpp (ed.), *The Philosophy of Karl Popper*, 2 vols., pp. 961–1197. La Salle, IL: Open Court.

Popper, Karl R. 1974/1992. *Unended Quest: An Intellectual Autobiography*. London: Routledge.

Popper, K. R. 1975/1981. "The Rationality of Scientific Revolutions," in I. Hacking (ed.), *Scientific Revolutions*, pp. 80–106. Oxford: Oxford University Press.

Popper, K. 1978. "Natural Selection the Emergence of Mind," *Dialectica*, 32:3–4, 339–355.

Portin, P. 2015. "The Development of Genetics in the Light of Thomas Kuhn's Theory of Scientific Revolutions," *Recent Advances in DNA and Gene Sequences* 9, 14–25.

Post, H. R. 1971. "Correspondence, Invariance and Heuristics: In Praise of Conservative Heuristics," *Studies in History and Philosophy of Science* 2:3, 213–255.

Psillos, Stathis. 1999. *Scientific Realism: How Science Tracks Truth*. London: Routledge.

Putnam, Hilary. 1978. *Meaning and the Moral Sciences*. Boston, London and Henley: Routledge & Kegan Paul.

Pyle, Andrew. 2000. "The Rationality of the Chemical Revolution," in Robert Nola and Howard Sankey (eds.), *After Popper, Kuhn, and Feyerabend*, pp. 99–124. Dordrecht: Kluwer Academic Publishers.

Quine, W. V. 1951. "Two Dogmas of Empiricism," *The Philosophical Review* 60:1, 20–43.

Quine, W. v. O. 1960. *Word and Object.* Cambridge, MA: The MIT Press.

Reichenbach, H. 1938/2006. *Experience and Prediction: An Analysis of the Foundations and the Structure of Knowledge.* Notre Dame, IN: University of Notre Dame Press.

Reingold, N. 1980. "Through Paradigm-Land to a Normal History of Science," *Social Studies of Science* 10:4, 475–496.

Reisch, G. 1991. "Did Kuhn Kill Logical Empiricism?," *Philosophy of Science* 58:2, 264–277.

Reisch, George A. 2005. *How the Cold War Transformed Philosophy of Science: To the Icy Slopes of Logic.* Cambridge: Cambridge University Press.

Reisch, G. A. 2016. "Aristotle in the Cold War: On the Origins of Thomas Kuhn's *Structure of Scientific Revolutions*," in R. J. Richards and L. Aston (eds.), *Kuhn's Structure of Scientific Revolutions at Fifty: Reflections on a Science Classic,* pp. 12–29. Chicago, IL: University of Chicago Press.

Reisch, George A. 2019. *The Politics of Paradigms: Thomas S. Kuhn, James B. Conant, and the Cold War "Struggle for Men's Minds."* Albany, NY: SUNY Press.

Renzi, B. G. 2009. "Kuhn's Evolutionary Epistemology and Its Being Undermined by Inadequate Biological Concepts," *Philosophy of Science* 76:2, 143–159.

Reydon, T. A. C., and P. Hoyningen-Huene. 2010. "Discussion: Kuhn's Evolutionary Analogy in *The Structure of Scientific Revolutions* and 'The Road since Structure," *Philosophy of Science* 77:3, 468–476.

Reynolds, A. 1999. "What Is Historicism?," *International Studies in the Philosophy of Science* 13:3, 275–287.

Richards, Robert J., and Lorraine Daston (eds). 2016. *Kuhn's Structure of Scientific Revolutions at Fifty: Reflections on a Science Classic.* Chicago, IL: University of Chicago Press.

Richardson, A. 2007. "'That Sort of Everyday Image of Logical Positivism': Thomas Kuhn and the Decline of Logical Empiricist Philosophy of Science," in A. Richardson and T. Uebel (eds.), *The Cambridge Companion to Logical Empiricism,* pp. 346–369. Cambridge: Cambridge University Press.

Richardson, A., and T. Uebel. 2007. "Introduction," in A. Richardson and T. Uebel (eds.), *The Cambridge Companion to Logical Empiricism,* pp. 1–10. Cambridge: Cambridge University Press.

Ritzer, G. 1981. "Paradigm Analysis in Sociology: Clarifying the Issues," *American Sociological Review* 46:2, 245–248.

RLE Undercurrents. 1997. "A Last, Loving Look at an MIT Landmark – Building 20," *RLE Undercurrents,* 9:2. www.rle.mit.edu/media/undercurrents/Vol9_2_Spring97.pdf (accessed May 24, 2021).

Rossiter, M. 1984. "The History and Philosophy of Science Program at the National Science Foundation," *Isis* 75:1, 95–104.

Rowbottom, Darrell P. 2019. *The Instrument of Science: Scientific Anti-realism Revitalized.* New York and London: Routledge, Taylor & Francis Group.

Rueger, A. 1996. "Risk and Diversification in Theory Choice," *Synthese* 109:2, 263–280.

Sankey, H. 2018a. "The Demise of the Incommensurability Thesis," in M. Mizrahi (ed.), *The Kuhnian Image of Science: Time for a Decisive Transformation?*, pp. 75–91. London: Rowman & Littlefield International Ltd.

Sankey, Howard. 2018b. "Kuhn, Relativism and Realism," in Juha Saatsi (ed.), *The Routledge Handbook of Scientific Realism*, pp. 72–83. London: Routledge.

Sauer, T., and R. Scholl (eds.). 2016. *The Philosophy of Historical Case Studies*, Boston Studies in the Philosophy and History of Science, Vol. 319. Dordrecht: Springer.

Scerri, Eric R. and Lee McIntyre. 1997. "The Case for the Philosophy of Chemistry," *Synthese* 111, 213–232.

Scheffler, Israel. 1967. *Science and Subjectivity.* Indianapolis, IN: The Bobbs-Merrill Company, Inc.

Schindler, S. 2018. *Theoretical Virtues in Science: Uncovering Reality Through Theory.* Cambridge: Cambridge University Press.

Schlick, M. 1930–1931/1959. "The Turning Point in Philosophy," in A. J. Ayer (ed.), *Logical Positivism*, pp. 53–59. New York, NY: The Free Press.

Schlick, Moritz. 1932–1933/1959. "Positivism and Realism," in A. J. Ayer (ed.), *Logical Positivism*, pp. 82–107. New York, NY: The Free Press.

Scholl, R. 2018. "Scenes from a Marriage: On the Confrontation Model of History and Philosophy of Science," *Journal of the Philosophy of History* 12:2, 212–238.

Schummer, Joachim. 2006. "The Philosophy of Chemistry: From Infancy Toward Maturity," in Davis Baird, Eric Scerri and Lee McIntyre (eds.), *Philosophy of Chemistry: Synthesis of a New Discipline*, pp. 19–39. Dordrecht: Springer.

Schuster, J. A. 2018. "The Pitfalls and Possibilities of Following Koyré: The Younger Tom Kuhn, 'Critical Historian,' on Tradition Dynamics and Big History," in R. Pisano, J. Agassi and D. Drozdova (eds.), *Hypotheses and Perspectives in the History and Philosophy of Science: Homage to Alexandre Koyré 1892–1964*, pp. 391–420. Dordrecht: Springer.

Science. 1883. "National Traits in Science," *Science* 2:35 (October 5, 1883), 455–457.

Sent, E.-M. 2004. "Behavioral Economics: How Psychology Made Its (Limited) Way Back into Economics," *History of Political Economy* 36:4, 735–760.

Sewell, W. H., Jr. 2005. *Logics of History: Social Theory and Social Transformation.* Chicago, IL: University of Chicago Press.

Shapere, D. 1964. "Review of *Structure of Scientific Revolutions*," *Philosophical Review* 73:3, 383–394.

Shapere, D. 1971. "Review: The Paradigm Concept," *Science* 172:3984 (May 14, 1971) 706–709.

Shapin, S. 1996. *The Scientific Revolution.* Chicago, IL: University of Chicago Press.

Shapin, Steven, and Simon Schaffer. 1985. *Leviathan and the Air-Pump: Hobbes, Boyle, and the Experimental Life*. Princeton, NJ: Princeton University Press.

Sharer, R. J., and W. Ashmore. 2003. *Archaeology: Discovering Our Past*, 3rd ed. New York, NY: McGraw-Hill.

Shearmur, J. 2017. "Popper's Influence on the Social Sciences," in L. McIntyre and A. Rosenberg (eds.), *The Routledge Companion to Philosophy of Social Science*, pp. 55–64. London: Routledge.

Shimony, A. 1979. "Paradigm Lost? A Review Symposium," *Isis* 70:253, 434–437.

Shirley, John W. 1951. "The Harvard Case Histories in Experimental Science: The Evolution of an Idea," *American Journal of Physics* 19, 419–423.

Smart, J. J. C. 1963. *Philosophy and Scientific Realism*. London: Routledge & Kegan Paul.

Staley, Richard. 2013. "Trajectories in the History and Historiography of Physics in the Twentieth Century," *History of Science* 51:2, 151–177.

Stanford, P. K. 2006. *Exceeding Our Grasp: Science, History, and the Problem of Unconceived Alternatives*. Oxford: Oxford University Press.

Stephens, J. 1973. "The Kuhnian Paradigm and Political Inquiry: An Appraisal," *American Journal of Political Science* 17:3, 467–488.

Stigler, G. 1963. "Letter to Thomas Kuhn – Dated 14 March 1963," in Thomas S. Kuhn Papers, MC240, Box 4: Folder 15, Structure of Scientific Revolutions, Correspondence S. Institute Archives and Special Collections, MIT Libraries, Cambridge, Massachusetts.

Strong, E. 1992. "Philosopher, Professor, and Berkeley Chancellor: Edward W. Strong. Interviews Conducted by Harriet Nathan in 1988." Berkeley: The Regents of the University of California. http://texts.cdlib.org/view?docId=kt2 p30025k&brand=calisphere&doc.view=entire_text (accessed July 30, 2020).

Tang, S. 2011. "Foundational Paradigms of Social Sciences," *Philosophy of the Social Sciences* 41:2, 211–249.

Thackray, Arnold, and Robert K. Merton. 1972. "On Discipline Building: The Paradoxes of George Sarton," *Isis* 63:4, 472–495.

Thagard, Paul. 1990. "The Conceptual Structure of the Chemical Revolution," *Philosophy of Science*, 57: 183–209.

Thorne, J. P. 1965. "Review of Paul Postal's *Constituent Structure: A Study of Contemporary Models of Syntactic Description*," *Journal of Linguistics* 1:1, 73–76.

Timmins, Adam. 2013. "Why Was Kuhn's *Structure* More Successful than Polanyi's *Personal Knowledge*," *HOPOS: The Journal of the International Society for the History of Philosophy of Science* 3: 306–317.

Toulmin, Stephen. 1970/1972. "Does the Distinction Between Normal and Revolutionary Science Hold Water?" in *Criticism and the Growth of Knowledge: Proceedings of the International Colloquium in the Philosophy of Science, London, 1965*, Vol. IV, reprinted with corrections, eds. Imre Lakatos and Alan Musgrave, pp. 39–47. Cambridge: Cambridge University Press.

Toulmin, Stephen. 1972. *Human Understanding*, Vol. 1. Princeton, NJ: Princeton University Press.

Trenn, Thaddeus. J. 1979. "Preface," in L. Fleck (ed.), *Genesis and Development of a Scientific Fact*, translated by Fred Bradley and Thaddeus J. Trenn, pp. xiii–xix. Chicago, IL: University of Chicago Press.

Truman, D. B. 1965. "Disillusion and Regeneration: The Quest for a Discipline," *American Political Science Review* 59:4, 865–873.

Tsou. J. Y. 2015. "Reconsidering the Carnap-Kuhn Connection," in W. J. Devlin and A. Bokulich (eds.), *Kuhn's Structure of Scientific Revolutions – 50 Years On*, pp. 51–69. Dordrecht: Springer.

van Fraassen, Bas C. 1980. *The Scientific Image*. Oxford: Clarendon Press.

van Fraassen, B. C. 1989. *Laws and Symmetry*. Oxford: Clarendon Press.

Vickers, Peter. 2018. "Historical Challenges to Realism," in Juha Saatsi (ed.), *The Routledge Handbook of Scientific Realism*, pp. 48–59. London: Routledge,

Walker, T. C. 2010. "The Perils of Paradigm Mentalities: Revisiting Kuhn, Lakatos, and Popper," *Perspectives on Politics* 8:2, 433–451.

Watkins, J. 1970. "Against 'Normal Science,'" in I. Lakatos and A. Musgrave (eds.), *Criticism and the Growth of Knowledge: Proceedings of the International Colloquium in the Philosophy of Science, 1965, Volume 4*, pp. 25–37. Cambridge: Cambridge University Press.

Weingart, S. 2015. "Finding the History and Philosophy of Science," *Erkenntnis* 80: 201–213.

White, H. 1963. "Review of Quantification: A History of the Meaning of Measurement in the Natural and Social Sciences, by Harry Woolf," *American Journal of Sociology* 69:1, 84–85.

Whitesides, George M. 2013. "Is the Focus on 'Molecules' Obsolete?" *Annual Review of Analytic Chemistry* 6: 1–29.

Wilkins, A. S. 1996. "Are There 'Kuhnian' Revolutions in Biology?" *BioEssays* 18:9, 695–696.

Worrall, John. 1989. "Structural Realism: The Best of Both Worlds?" *Dialectica* 43:1–2, 99–124.

Worrall, John. 2002. "Normal Science and Dogmatism, Paradigms and Progress: Kuhn 'versus' Popper and Lakatos," in T. Nickles (ed.), *Thomas Kuhn*, pp. 65–100. Cambridge: Cambridge University Press.

Wray, K. B. 2002. "Social Selection, Agents' Intentions, and Functional Explanation," *Analyse & Kritik* 24: 72–86.

Wray, K. B. 2003. "Is Science Really a Young Man's Game?" *Social Studies of Science* 33:1, 137–149.

Wray, K. B. 2010. "Philosophy of Science: What Are the Key Journals in the Field?" *Erkenntnis*, 72: 423–430.

Wray, K. B. 2011. *Kuhn's Evolutionary Social Epistemology*. Cambridge: Cambridge University Press.

Wray, K. Brad. 2015a. "Pessimistic Inductions: Four Varieties," *International Studies in the Philosophy of Science* 29:1, 61–73.

Wray, K. Brad. 2015b. "The Methodological Defense of Realism Scrutinized," *Studies in History and Philosophy of Science* 54, 74–79.

Wray, K. Brad. 2015c. "Kuhn's Social Epistemology and the Sociology of Science," in W. J. Devlin and A. Bokulich (eds.), *Kuhn's Structure of Scientific Revolutions – 50 Years On*, pp. 167–183. Dordrecht: Springer.

Wray, K. Brad. 2016. "The Influence of James B. Conant on Kuhn's *Structure of Scientific Revolutions*," *HOPOS: The Journal of the International Society for the History of Philosophy of Science*, 6:1, 1–23. DOI: https://doi.org/10.1086/685542

Wray, K. B. 2017. "Kuhn's Influence on the Social Sciences," in L. McIntyre and A. Rosenberg (eds.), *The Routledge Companion to the Philosophy of Social Science*, pp. 65–75. London and New York: Routledge.

Wray, K. Brad. 2018a. *Resisting Scientific Realism*. Cambridge: Cambridge University Press.

Wray, K. Brad. 2018b. "Thomas Kuhn and the T. S. Kuhn Archives at MIT," *OUPblog*, May 7, 2018. https://blog.oup.com/2018/05/thomas-kuhn-archives-mit/ (accessed June 5, 2021).

Wray, K. B. 2019. "Discarded Theories: The Role of Changing Interests," *Synthese* 196, 553–569.

Wray, K. B. 2020. "Paradigms in Structure: Finally, a Count," *Scientometrics* 125, 823–828.

Zammito, J. H. 2004. *A Nice Derangement of Epistemes: Post-Positivism in the Study of Science from Quine to Latour*. Chicago, IL: University of Chicago Press.

Zuckerman, H. 1977/1996. *Scientific Elite: Nobel Laureates in the United States*. Piscataway, NJ: Transaction Publishers.

Archival Sources

Thomas S. Kuhn Papers, MC240, Box 20: Folder 12, NSF Research Reports.

Index

Andersen, Hanne, 11, 27, 29
anomaly/anomalous, 19, 23, 37–38, 48,
    51–52, 70, 85, 86, 97, 111, 127, 138, 139, 143,
    145, 147, 159
anthropology, 84–85, 94, 100–102, 137
Aristotle experience, the. *See* Kuhn's education/
    career
astronomy, 60, 63, 91, *See also* revolutions in
    science: Copernican Revolution in
    astronomy, the
atom, 38, 72, 124, 131, 175, 183, 192
atomic bomb, 26–27
atomic theory, 57–59, 98, 161, 175

Barnes, Barry, 45, 110–111, 114, 115, 116–117,
    184–185
biology, 150, 195
Bird, Alexander, 11, 67, 68, 146, 150–151
*Black-Body Theory and the Quantum
    Discontinuity, 1894–1912*, 63, 125–132, 138,
    163, 200
Bloor, David, 45, 109, 110, 116
Bohr, Niels, 4, 123–124, 131, 175,
    176–177, 180
Bruner, Jerome, 38, 52, 85, 139, 203
Bush, Vannevar, 27, 28

Carnap, Rudolf, 60, 65–68, 70, 72–73, 168, 175,
    177–178
case study method, the, 17, 57, 171–172, *See also*
    Harvard College/University: Harvard Case
    Histories in Experimental Science, The
Cavell, Stanley, 4, 69, 73, 134
chemistry, 31, 38, 42, 47, 159
Cohen, I. B., 6, 74
Cold War, the, 7, 26, 62
Cole, Jonathan R., 105–107, 147
community, scientific, 22, 29, 35, 37, 38, 45,
    47–49, 70, 90, 93, 103, 106, 109–110, 112, 117,
    131, 147, 160, 165, 166, 188–189, 197
Conant, James, 5, 200, 204

Conant, James B., 1, 3, 4, 5, 9–12, 17, 21, 23, 25–49,
    50, 56–59, 64, 69, 74, 133, 141, 147, 155, 160,
    200, 203
conceptual scheme, 21, 34–35, 37, 40–42, 44,
    56, 110
confirmation/verification, 20, 36, 51, 69, 70, 78,
    142, 162, 164, 177
context of discovery/context of justification,
    74, 162
Copernican Revolution, The, 12, 35, 41–42, 56, 59,
    85, 130, 132, 161, 203
crisis, 35, 38, 44, 47, 52–53, 55, 76, 86–87, 96, 97,
    109, 113, 126, 143, 182, 184, 197

Dalton, John/Daltonian, 53–54, 58–59, 98, 170
Darwin, Charles/Darwinian, 99, 137, 189
Daston, Lorraine, 136, 139–140, 144, 145, 147, 149
disciplinary matrix, 92–93, 163
discovery, scientific, 18, 42, 43, 51–53, 71, 77, 105,
    127, 130, 138, 145–147, 158, 196
Doppelt, Gerald, 170, 190–191
duck/rabbit image, 13, 17, *See also* Gestalt
    psychology/figure
Duhem, Pierre, 60–61, 176
Durkheim, Émile, 93, 144

economics, 6, 42, 84–85, 88–89, 91, 94–95, 97,
    98–100, 137, 148, 196
Edge, David, 29, 108–109
Edmonds, David, 59–60, 67, 157, 168
Einstein, Albert, 37, 51, 125, 131, 176, 178, 181
entities, theoretical, 38, 175–176, 177, 179, 180,
    190–191, 192
epistemology, 46, 107, 117–118, 192
    naturalized, 68, 172, 203
    social, 26, 48, 116
Essential Tension, The, 12, 74, 107, 156, 166, 200
essential tension in science, the, 27, 143
evolution, biological, 22–23, 70, 99, 145, 169, 205
exemplar, 20, 39, 43, 44, 55, 78–79, 86–87, 92–93,
    162–163, 200–201